T0305748

Lecture Notes in
Water Policy

World Scientific Lecture Notes in Economics and Policy

ISSN: 2630-4872

Series Editor: Ariel Dinar *(University of California, Riverside, USA)*

The World Scientific Lecture Notes in Economics and Policy series is aimed to produce lecture note texts for a wide range of economics disciplines, both theoretical and applied at the undergraduate and graduate levels. Contributors to the series are highly ranked and experienced professors of economics who see in publication of their lectures a mission to disseminate the teaching of economics in an affordable manner to students and other readers interested in enriching their knowledge of economic topics. The series was formerly titled World Scientific Lecture Notes in Economics.

Published:

Vol. 11: *Lecture Notes in Water Policy*
by David Feldman

Vol. 10: *Lecture Notes in International Trade Theory: Classical Trade and Applications*
by Larry S. Karp

Vol. 9: *Lecture Notes in International Trade: An Undergraduate Course*
by Priyaranjan Jha

Vol. 8: *Lecture Notes in State and Local Public Finance: (Parts I and II)*
by John Yinger

Vol. 7: *Economics, Game Theory and International Environmental Agreements: The Ca' Foscari Lectures*
by Henry Tulkens

Vol. 6: *Modeling Strategic Behavior: A Graduate Introduction to Game Theory and Mechanism Design*
by George J. Mailath

Vol. 5: *Lecture Notes in Public Budgeting and Financial Management*
by William Duncombe

For the complete list of volumes in this series, please visit
www.worldscientific.com/series/wslnep

World Scientific Lecture Notes in Economics and Policy – Vol. 11

Lecture Notes in Water Policy

David L Feldman
University of California, Irvine, USA

World Scientific
NEW JERSEY • LONDON • SINGAPORE • BEIJING • SHANGHAI • HONG KONG • TAIPEI • CHENNAI • TOKYO

Published by

World Scientific Publishing Co. Pte. Ltd.
5 Toh Tuck Link, Singapore 596224
USA office: 27 Warren Street, Suite 401-402, Hackensack, NJ 07601
UK office: 57 Shelton Street, Covent Garden, London WC2H 9HE

Library of Congress Control Number: 2021054574

British Library Cataloguing-in-Publication Data
A catalogue record for this book is available from the British Library.

World Scientific Lecture Notes in Economics and Policy — Vol. 11
LECTURE NOTES IN WATER POLICY

ISBN 978-981-124-223-6 (hardcover)
ISBN 978-981-124-318-9 (paperback)
ISBN 978-981-124-224-3 (ebook for institutions)
ISBN 978-981-124-225-0 (ebook for individuals)

For any available supplementary material, please visit
https://www.worldscientific.com/worldscibooks/10.1142/12417#t=suppl

Desk Editors: Aanand Jayaraman/Sylvia Koh

Typeset by Stallion Press
Email: enquiries@stallionpress.com

Printed in Singapore

About the Author

David Lewis Feldman is Professor of Urban Planning & Public Policy and Political Science at UC Irvine. He also serves as director of the Master's of Public Policy program and as director of Water UCI. The latter currently undertakes research on various water issues in California and internationally, as well as producing education and outreach activities for decision-makers and the public. Feldman's PhD is from the University of Missouri, and B.A. is from Kent State University. He served as lead author for a U.S. Climate Change Science Program report on climate and water; has served on external advisory boards on two NSF-funded research centers on environmental issues in the western U.S.; and was co-Principal Investigator on an NSF-Partnerships for International Research and Education project with Australian universities on water reuse. He has also collaborated on research projects in Israel and the European Union. The author/co-author of nearly 100 articles and book chapters, he has most recently written: *The Governance of Water Innovations* (forthcoming, 2022); *Water Politics — Governing Our Most Precious Resource* (Polity, 2017); and *The Water Sustainable City* (Elgar, 2017).

Contents

About the Author v

Introduction ix

Section 1 Introduction — Freshwater Management as Global Policy Challenge 1

Section 2 Past as Prologue — Water Resources Policy in Historical Perspective 25

Section 3 US Policies in a Global Context — Law, Practice, and Institutions 55

Section 4 Who Controls Freshwater? Allocation and Uses 77

Section 5 Water Quality — Impacts to Health, Environment, and Well-Being 91

Section 6 Sources of Water Conflicts — Diversion, Depletion, and Degradation 117

Section 7 California as "Hydraulic Empire" — Fact, Fiction, Fantasy? 151

Section 8 California's Water — Problems and Solutions in Global Context 177

Section 9 Ethics, Values, and Water — The Challenges of Adaptive Management 199

Index 221

Introduction

Freshwater is our planet's most precious resource — essential for life itself. Despite this fact, many people across our planet face difficulty finding safe, clean, potable water. A US State Department report contends that the world's thirst for water may become a human security crisis by 2040. The World Bank reports many developing nations face catastrophe from intensive irrigation, urbanization, and deteriorating infrastructure; and, numerous reports contend that in many places un-treated wastewater is still released directly into the environment. This is particularly true in low-income countries, which on average treat less than 10% of their wastewater discharges.

In short, we face three imminent challenges regarding freshwater: (1) demands by agriculture, cities, industry, and energy production are increasing; (2) severe pollution from various contaminants, and growing withdrawals are limiting the capacity of waterways to dilute contaminants — threatening human and aquatic life; and (3) climate change will cause periods of frequent and severe droughts — punctuated by acute periods of flooding.

The goal of this course is to illuminate how the governance of freshwater is a political, social, economic, cultural, and ecological challenge. The management and provision of water are not merely technical problems whose resolution hinges on hydrological principle, cost, or engineering feasibility. They are products of decisions made by governments, businesses, and interest groups that exercise control over who has access to water, how they use it, and in what condition they receive it. We will discuss basic knowledge about water supply and quality; the evolution of water policy in different societies; the importance of water to human and environmental health; the role of law, politics, and markets in its allocation, use, and protection; and the importance of ethics in its equitable provision.

We will focus on the global competition for water and the impacts of this competition on conflicts over supply and quality — in places as diverse as the Middle East, Europe, Africa, Latin America, and Asia. While disputes over water are found everywhere, we will also pay special attention to the US, particularly the West and California — where water has long been a focal point of contention. My objectives in this course are three-fold:

- To enable you to understand and analyze water problems and think creatively about possible solutions. Here in Southern California, many policy innovations are being introduced that have world-wide application, including conservation and innovative pricing systems, reclaiming of waste-water and re-use of stormwater, and de-salination, among others.
- To help you understand the importance of different fields of study to water resource policy — including social ecology, urban studies, social sciences, bioscience, public health, earth systems science, history, philosophy, the humanities, and engineering (among others). Sound freshwater management requires a broad knowledge base and sensitivity to diverse needs.
- To help you apply what you learn in the course to improve water management in your own backyard: this includes your local watershed, and even your household. Stewardship over this precious resource begins with personal awareness, a willingness to change one's behavior, and a commitment to life-long service to one's community.

Water politics comprises more than the actions of government. It embraces the activities of private businesses that treat, distribute and sell water for a profit; civil society groups that avidly defend the rights of people who demand access to affordable water; and entire nations that covet and compete for shared river- or groundwater basins. We employ a three-fold structure for understanding water politics: the process of decision-making, the exercise of power, and the purposes governance aims to achieve.

Process

A key to understanding water politics is the process by which issues get the attention of officials (i.e., what political scientists call agenda setting), and the means by which policies are formulated and applied through law, rule, treaty, or common everyday practice. Process encompasses negotiation, bargaining, and accommodation among various interests that use, manage, and provide water and includes such entities as corporations; scientific, legal, and other experts; large user groups (e.g., farmers, industrial and commercial sectors, urban utilities); as well as environmental groups and citizen organizations. Individual decision- makers holding formal positions are also involved.

Power

All water politics involves power. While "power" sometimes connotes manipulation, cajoling, or undue pressure exerted by a few individuals over decisions, these are only the most visible examples of its exercise in water politics. Strategically-positioned groups are able to successfully advocate for water policies because they own or control capital needed for investing in water infrastructure, possess legally enshrined rights to water, have "spiritually- endowed" authority in its management — or, lastly — are able to physically coerce others. While power has had an important role in shaping water politics in modern industrialized countries and developing countries alike during our present era, its importance has also been recognized in accounts of water decision- making in the distant past. Many scholars have chronicled connections between power, interest group dynamics, decision- maker access, and coalition building for generations.

Purpose

Water politics is a purposeful activity. Participants seek tangible objectives such as ensuring an adequate water supply of a quality sufficient for potable use; guaranteeing that lakes and rivers are safe for Water politics is a purposeful activity. Participants seek tangible objectives such as ensuring an adequate water supply of a quality sufficient for potable use; guaranteeing that lakes and rivers are safe for fishing, and protecting its available quantity to serve a variety of societal needs. Water politics participants hold conflicting beliefs about the value of water, its most and beneficial uses, and what constitutes an appropriate means of acquiring it. These conflicting beliefs manifest themselves in major differences over issues. For example: what objectives should the building of a reservoir achieve? How much should users be charged for water — and how should these charges be levied — e.g., by variation in volumes used? And do the health and ecological risks of pollution from some economic activity outweigh the societal benefits of using water to generate energy, manufacture goods, of produce food for human consumption?

Section 1

Introduction — Freshwater Management as Global Policy Challenge

How Could This Happen?

- Fall 2014: discovery that corrosive water leached lead from century old galvanized iron-pipes.
- Local, state, US officials tried to reassure residents that:
  - There was no health problem.
  - They used anti-corrosive phosphates to combat pipe corrosion and leaching of lead (they didn't).
- Crisis could have been averted if precautions taken — lack of funding was initial excuse. State now providing $600 million in aid (2020).

Some Key Issues in Water Policy

- *Water stress* — defined as an imbalance between supply and demand, it's caused by climate change, growing demands, societal inequities.
- *Pollution* — threats to water quality are growing; some are caused by legacy problems, others by newer forms of pollutants. All threaten human and ecological health.
- *Competition* — worldwide, many sources of water are shared among countries; this causes political disputes within and between nations.*
- *Remedies are problems too* — newer measures to provide additional supplies (e.g., desalination, wastewater reuse) — may generate public concerns.

Understanding Water Stress

Global water demands and trends

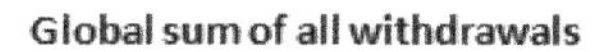

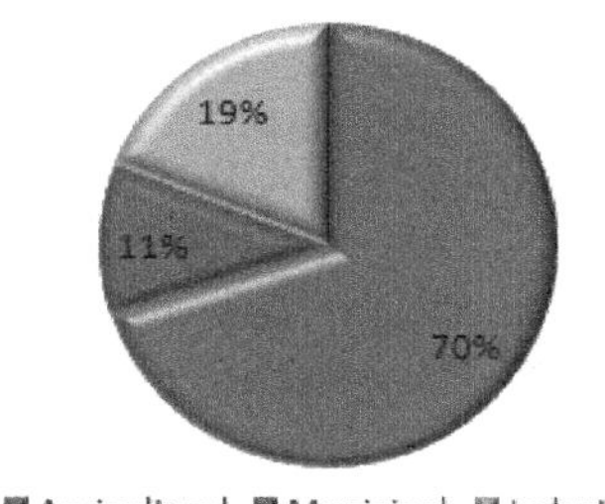

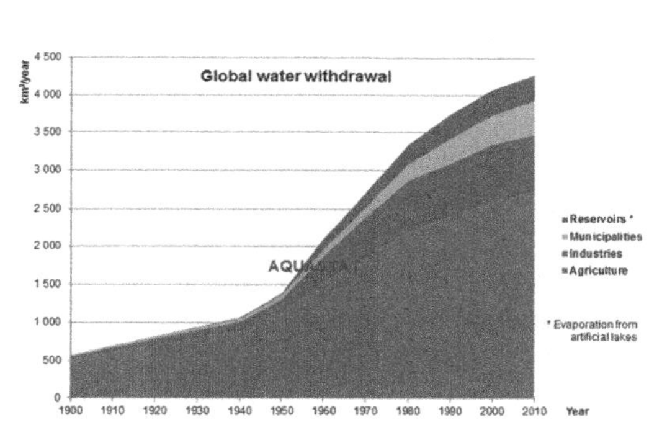

- In 20th century *global demands* rose six-fold, more than twice the rate of population growth, due to food and fiber production.
- *Food and fiber* demands account for 2/3 of global water use.

Sources: UN- FAO AQUASTAT database, 2012 and World Meteorological Organization.

*More than 260 river basins, home to over three billion people, are shared by two or more countries.

Disparities in Water Demand — By Country

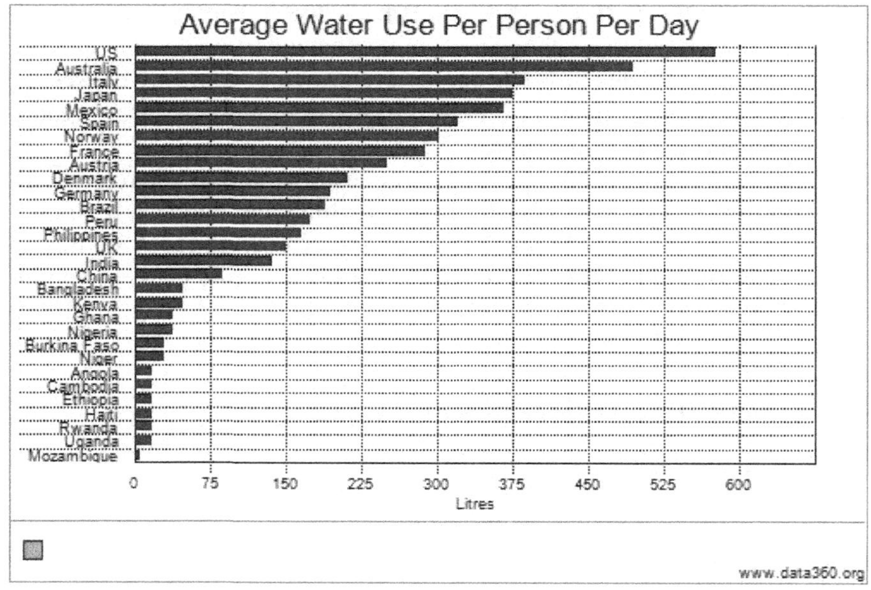

04/15/2010, James Shen: http://hdr.undp.org/en/media/HDR06-complete.pdf

- Developed countries use more water per person per day due to affluence — food and energy demands and *available infrastructure to deliver and treat water.*
- Poses *environmental justice* implications.

And Climate Change?

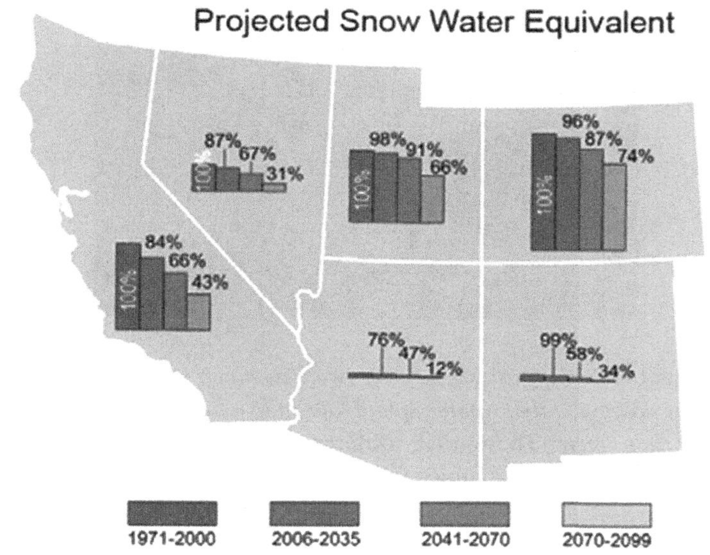

Source: US Global Change Research Program (2018).

Water Stress — Colorado River

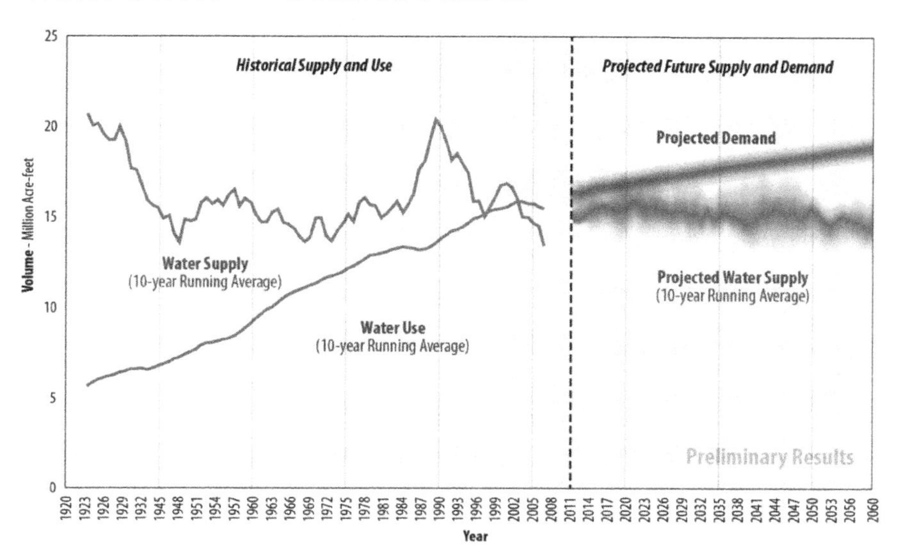

Figure depicting volume (annual flow) and demand (US BuRec, 2012).

Societal Inequities — Water as an Environmental Justice Problem

"In parts of Africa, *girls spend over twice as much time fetching water daily as do boys; adult women spend four to five times longer collecting water than men.* These time differentials affect opportunities for schooling ... and opportunities for pursuing economic opportunities"

— UN Water and Gender Report, 2014.

Women bear the main responsibility for collecting water in sub-Saharan Africa

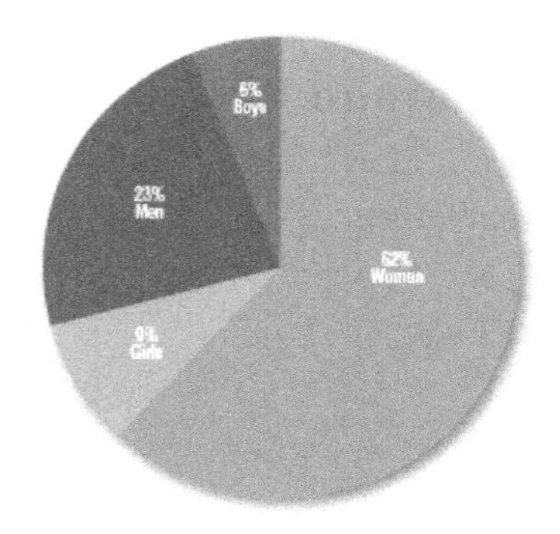

FIGURE 37 Distribution of the water collection burden among women, children under age 15 and men in households without piped water on premises, 25 countries in sub-Saharan Africa, 2006-2009 (per cent)

Source: MICS and DHS surveys from 25 sub-Saharan African countries

"We must work to ensure that no child dies from a preventable water-related disease, that no girl fears going to school for lack of access to a separate toilet, that no woman walks six kilometers to collect water for her family, and that no war is ever fought over water".

— Under Secretary for Civilian Security, Democracy, and Human Rights Maria Otero, March 22, 2011.

Gender Equity and Water — Mexico

- *Keys to reform include*:
  - Reforming land tenure systems to ensure women can inherit land, obtain collateral.
  - Promoting women's participation in irrigation collective meetings, leadership roles.

Other Inequities ... Debate Over Water as a Human Right

- Attacked in the street by strangers. Death threats over the phone. Dragged through court over bogus legal challenges.
- These are some challenges faced by Rodrigo Mundaca, who fights for basic water needs.
- Rodrigo's country of Chile is one place where access to water has been privatized.

Cape Town, South Africa — March 2018: Queuing for water.

By late 2019, as a result of rainfall return, aggressive conservation, water rationing, city is gradually recovering.

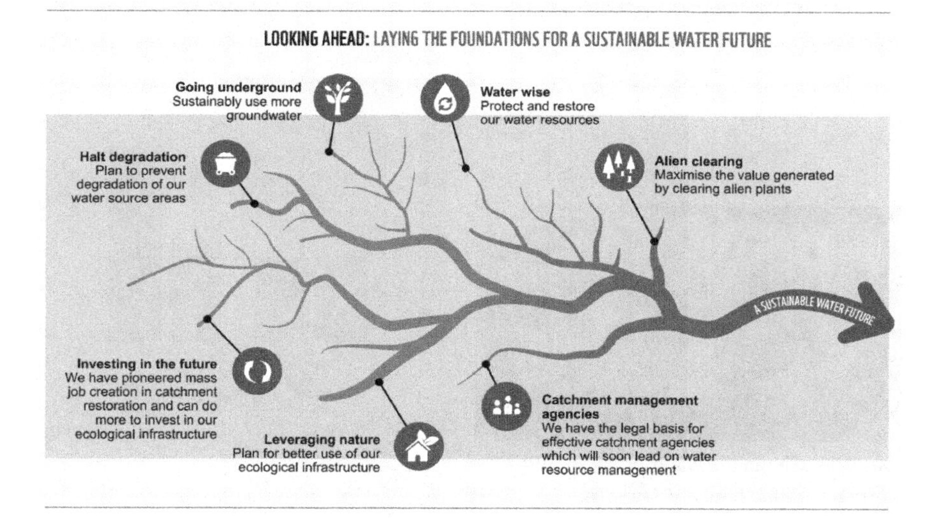

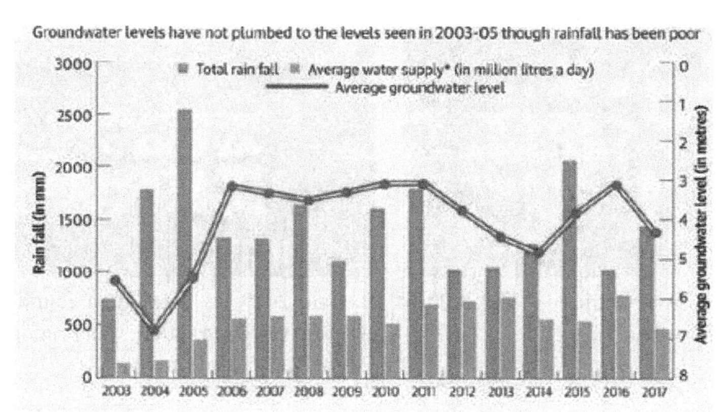

By 2030 the demand for water in India will be twice as much as the supply

Demand will be 101% of supply

Year

Supply Demand

Scroll.in

Source: NITI Aayog

See: Why Chennai's water crisis should worry you, *The Times of India*, June 21, 2019.

https://timesofindia.indiatimes.com/city/chennai/why-chennais-water-crisis-should-worry-you/articleshow/69899842.cms

- *Sources of problem*?
  - Protracted and serious drought.
  - Over-drafting of local groundwater with few regulations.
  - Poor urban planning — government encouraged growth of an IT sector in and around city; little thought given to supply.
  - Water now delivered by tankers filled with supplies purchased cheaply from poor local farmers.

For Discussion

- Are these problems human made or natural?
- What human activities contribute to these problems?
- Who is most affected by these problem?

Pollution and Threats to Water Quality

Human Health and Water Quality — Sanitation

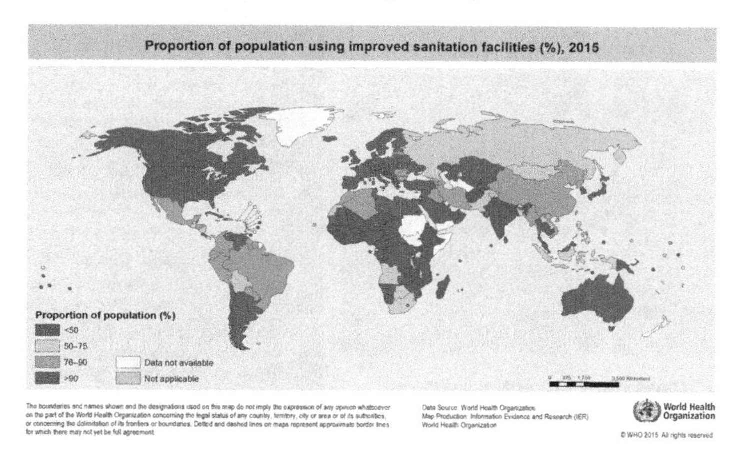

Water Quality, Climate Variability, Nutrient Pollution — a Growing Problem

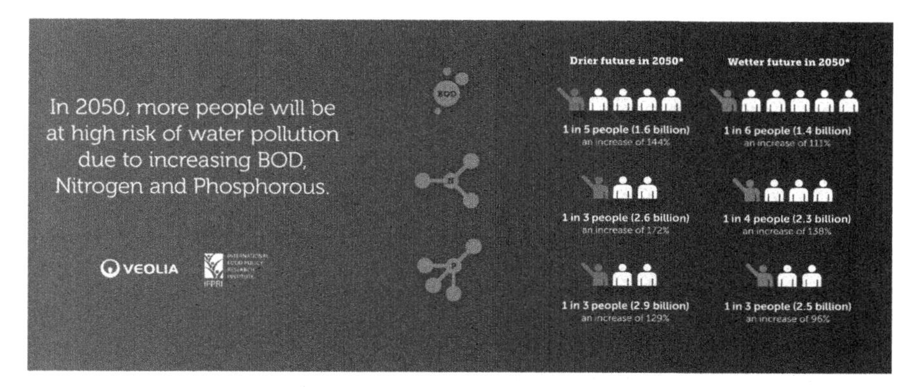

- *BOD = Biological oxygen demand*: amount of oxygen needed for microbial metabolism of organic compounds in water:
  - *Higher* BOD = poor quality; lower = better capacity to sustain aquatic life.
  - Greatest contributor to high BOD? runoff from agricultural fertilizers.

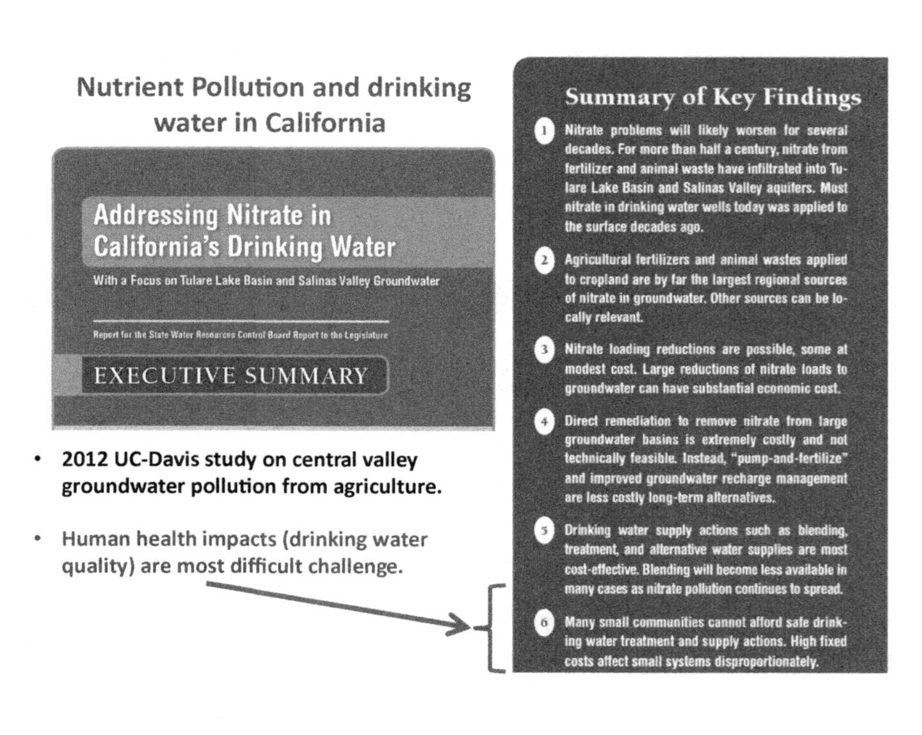

Examples of Nutrient Pollution

Algae blooms caused by fertilizer runoff — Taihu Lake, Jiangsu province, China.

Algae kills fish, contaminates drinking water.

Water Quality Problems can Stem from Unusual Sources

What about rainwater?

- Fine for watering ornamental plants, but...
- Test for *E. coli* if it is used on edible plants.
- Roof run-off:
  - Climate
  - Age of roof
  - Materials (metal?)
  - Air quality
  - Slope of roof
  - Temperature

Los Angeles River outflow

Urban runoff is leading cause of water pollution in Southern California:

- In 2014 — US Supreme Court held that Los Angeles County is responsible for untreated stormwater pollution that plagues waterways.
- In 2018 — LA county voters chose a property parcel surtax to provide measures to avert/store storm-water runoff, provide new sources of supply.

Competition Over Shared Freshwater

INTELLIGENCE COMMUNITY ASSESSMENT
ICA 2012-08, 2 February 2012
This is an IC-coordinated paper.

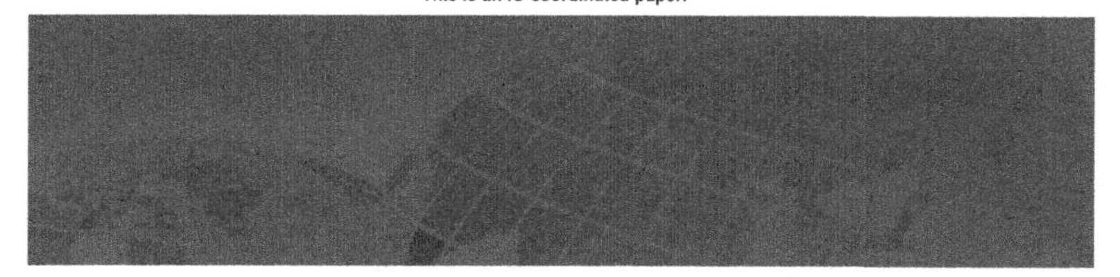

This report — requested by the Department of State — is designed to answer the question: How will water problems (shortages, poor water quality, or floods) impact US national security interests over the next 30 years? We selected 2040 as the endpoint of our research to consider longer-term impacts from growing populations, climate change, and continued economic development. However, we sometimes cite specific time frames (e.g., 2030, 2025) when reporting is based on these key dates.

- Commissioned by former US Secretary of State Hillary Rodham Clinton.
- By 2040, flooding, drought, poor water quality will create instability and regional tension; hinder ability to grow food, produce energy.
    - Greatest challenges? North Africa, Mid-East, South Asia.
    - Wars less likely than are chronic tensions.
    - Improved water saving/reclaiming technologies can address some, but NOT all problems — *agreements between countries will be required to solve problems.*

Some Transboundary Water Conflicts

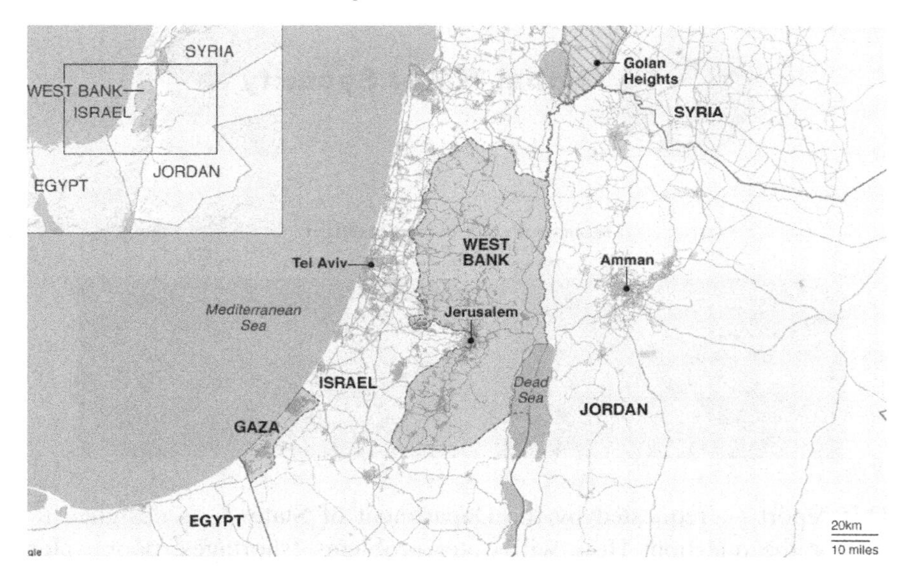

- Syria and Israel — frequent, violent conflict over control of upper Jordan River basin and tributaries (1950s–1960s).
- Israel and Jordan — lower Jordan basin and Dead Sea flow depletion issues (1960s–present).
- Palestine, Israel, Jordan — allocation of water supply from Jordan basin and groundwater basins.
- Palestine (including Gaza) and Israel — over sharing and protection of water infrastructure.

Sharing Water? — A Middle East Example

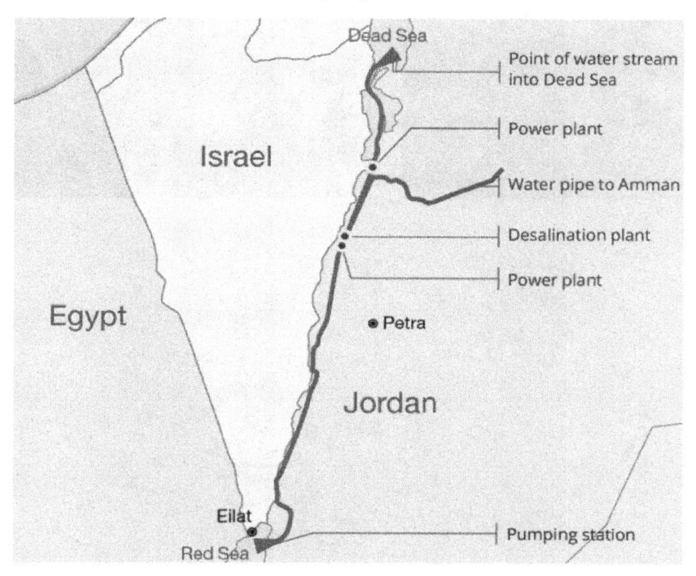

- Proposed *Red Sea-Dead Sea canal/desalination project* would pump water from Red Sea to desalination plant in Jordan.
- Desalinated water goes to Amman; brackish water to Dead Sea to halt shrinkage.
- Estimated cost = $10 billion ($2.7 billion from World Bank).
- Currently part of negotiations over a renewed *Jordan–Israel Peace Treaty* (1994) which MAY not be renewed.
- Some Israelis feel project is not economically beneficial to Israel; some Jordanians believe Israel cannot be trusted to address other "inter-linked" water issues.
- Countries would rather cooperate than fight over water — however, that does NOT mean there won't be *low-intensity conflict.*

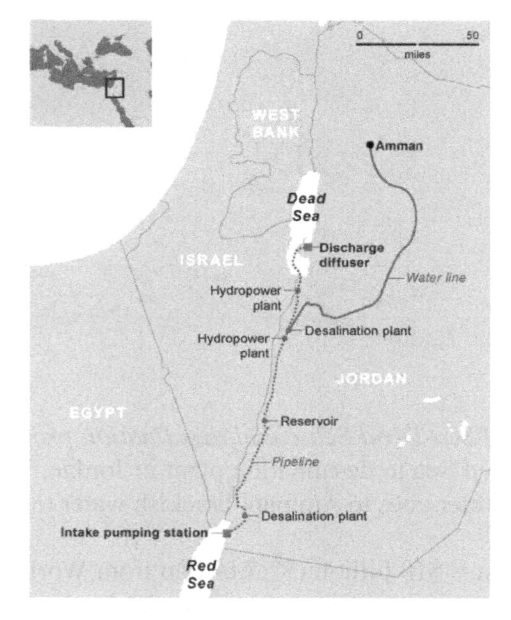

"Israel Mulls Advancing Red Sea-Dead Sea Canal Project to Mend Relations With Jordan".

Yaniv Kubovich and Noa Landau November 3, 2018 Haaretz.

The canal is one of several ideas that have been proposed in advance of the expected negotiations with Amman on the two annexes in the Israel–Jordan peace treaty, which King Abdullah recently announced will not be renewed.

Novel Remedies are Problems, too

Should we Desalinate?

- Poseidon Resources completed a $950 million plant near Carlsbad in 2016.
- Produces 50 million gallons/day; serves 300,000 people.
- *Project issues:*
    - Cost: $2000/acre-foot.
    - Energy use: 33 MW; enough for 80,000 homes.
    - Ecological impact: brine disposal impacts on marine life?
- Other issues:
    - If this is your primary water source, what do you do if it has to be shut-down?
    - With population growth occurring inland, we'd need to pump water up-hill — more energy needed.
    - Globally, desalination is popular in regions with high water demand, abundant energy supplies, few alternatives — e.g., Israel, Gulf states.

ROBERT GAUTHIER Los Angeles Times

THE TOWN of Marina, Calif., on Monterey Bay has long hosted industrial activity, like the Cemex sand mine, which has sat on the coastline for a century. Now, many in the town oppose a new desalination plant.

A WAR OVER WATER

Environmental justice has come to the fore as the coastal commission weighs a desalination plant on Monterey Bay

- What's at issue here?
  - Proposal for a desalination plant arose because Monterey peninsula lacks water to support continued growth.
  - Problem? Residents of Marina:
    - Won't receive any of the water.
    - Will end up partly paying for it in higher rates.
    - Will have an unsightly industrial facility in their "backyard".
    - Believe they are victimized by being an under-represented, minority, and low-income community.

Should We Re-Use Wastewater?

Can Conservation Work? — Los Angeles as Example

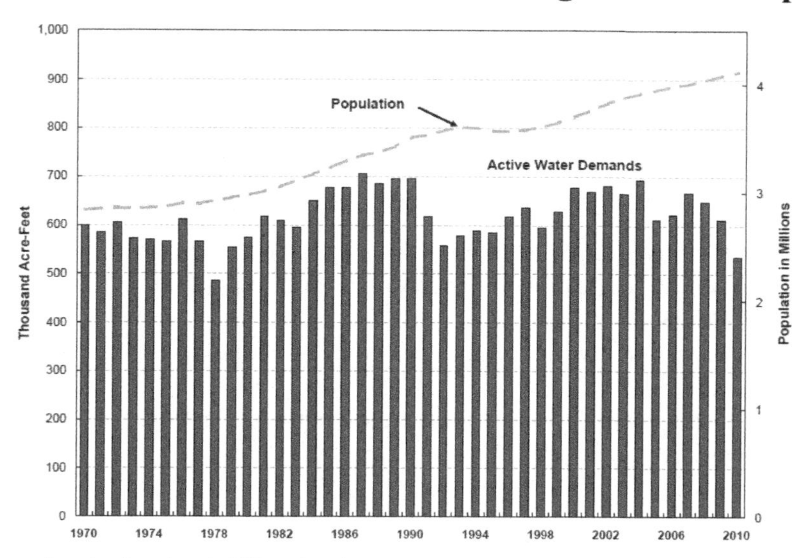

Source: Securing Los Angeles' Water Supply — LADWP — Urban Water Management Plan — Los Angeles, 2010 — data includes entire Los Angeles region served by Department of Water and Power.

- Los Angeles is a water stressed region.
- Creative use of rate structures, public outreach, replacing of appliances with water efficient models all reduced *per capita use*.
- Net result? City grew by over 1 million people in 40 years but water use declined.
- *This remedy* was both technically feasible AND publicly acceptable.

California Urban Water Demands

PER CAPITA URBAN WATER USE HAS BEEN FALLING AND FELL STEEPLY DURING THE LATEST DROUGHT

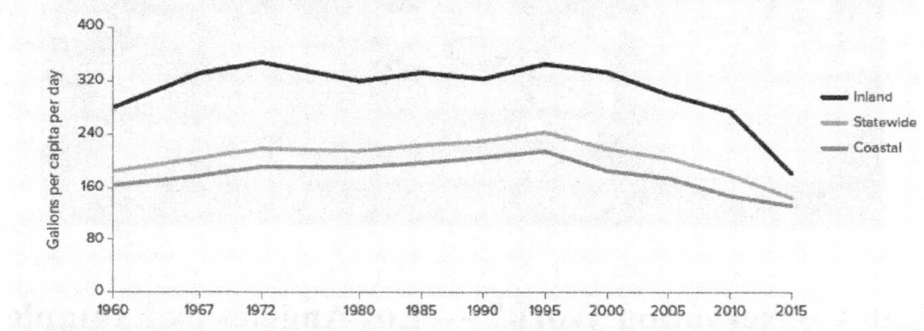

SOURCE: Author calculations using data from the California Department of Water Resources, *California Water Plan Update* (various years).

NOTES: The figure shows "applied" water delivered to homes and businesses. "Net" water use—i.e., the volume consumed by people or plants, embodied in manufactured goods, evaporated, or discharged to saline waters—is lower. The totals exclude water used by power plants and groundwater recharge projects and water lost during conveyance. Except for 2015 (a severe drought year), the estimates are for normal or "normalized" rainfall years (i.e., adjusted to levels that would have been used in a year of normal rainfall). Estimates are for water years (October to September). Inland areas tend to have higher per capita use because of higher temperatures and larger landscaped areas.

Source: *California's Water* — Public Policy Institute of California, November 2018.

What Makes a Water Remedy Acceptable?

- *Technical feasibility* — science and engineering support it.
- *Economical and fair* — affordable and equitable compared to alternatives.
- *Environmental impact and risk low* — adverse effects can be mitigated.
- *Public acceptability* — public trusts, and has a voice in shaping, options.

Note: (A) substitution; (B) regeneration; (C) reduction.

Leadership is one key to making remedies acceptable

Governor Brown Declares Drought State of Emergency
January 17, 2014

SAN FRANCISCO — Governor Brown announces Drought State of Emergency with Natural Resources Agency Secretary John Laird, DWR

Director Mark Cowin, Water Resources Control Board Chair Felicia Marcus, Office of Emergency Services Director Mark Ghilarducci.

- April 1, 2015 — Governor issues Executive Order directing State Water Resources Board to *impose restrictions on water suppliers to achieve a statewide 25% reduction in potable urban use; require commercial, industrial, institutional users to implement water efficiency measures; prohibit irrigation with potable water of ornamental turf in public street medians; prohibit irrigation with potable water outside newer homes and buildings that is not delivered by drip or micro-spray systems.*

Conclusions

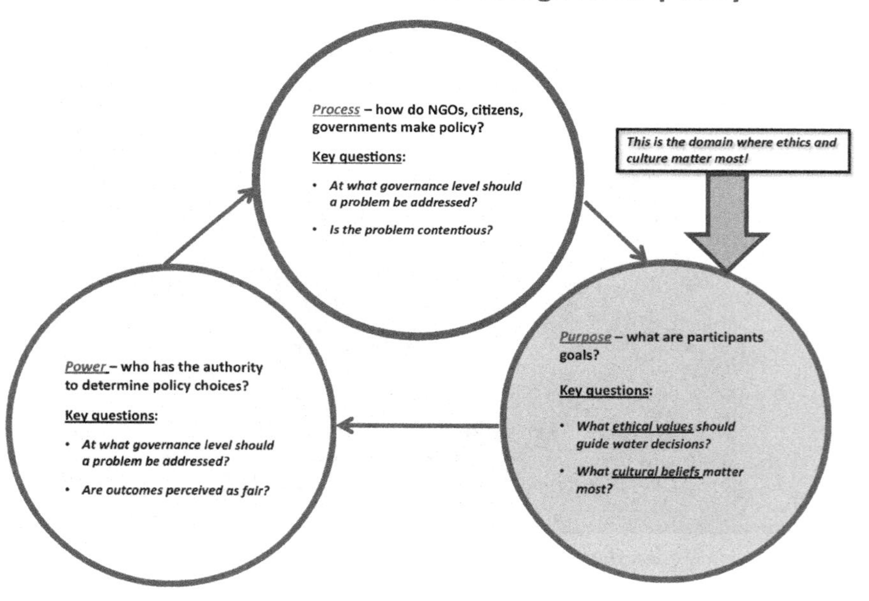

Section 2

Past as Prologue — Water Resources Policy in Historical Perspective

China — Water Innovations and Imperial Control

Dujiangyan irrigation project — Qin dynasty 256 BC — Min River, Sichuan Province.

图 4-85 水转翻车
（引自（明）宋应星《天工开物》）

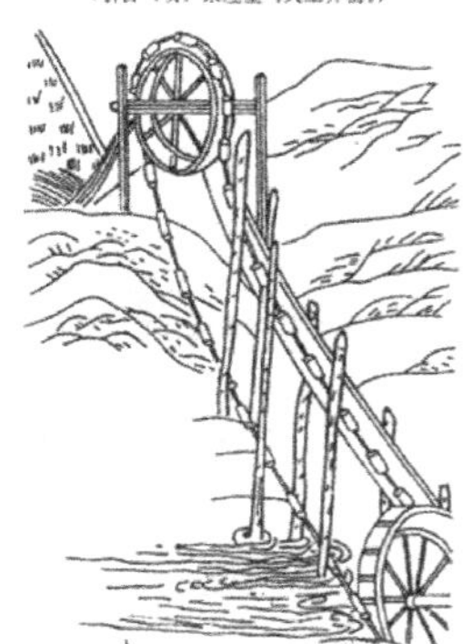

Tiangong Kaiwu Chain Pumps, Ming dynasty (c. 1300 — 1400 AD).

Note the impoundment structure at top of the picture which can store floodwaters or release them as the situation requires.

- PURPOSE: Water provision a major source of technological innovation.
- POWER/PROCESS: Allocation, navigation, flood control were centrally managed.

Irrigation and the Rise of Mesopotamia

- Mesopotamia harnessed Tigris and Euphrates rivers to:
  - Supply clean, reliable water.
  - Allow crops to be grown where water was not naturally available.
  - Enable rivers to be utilized to their full potential.
  - Enable more people to move from hunter-gatherers to farming and urban residents.
  - Flooding eventually filled irrigation canals with silt and destroyed irrigation systems.

Irrigation made it possible for Mesopotamians to grow food in places not naturally having water, such as the Tigris Valley shown below.

Petra, Jordan — World's Oldest Water Supply System?

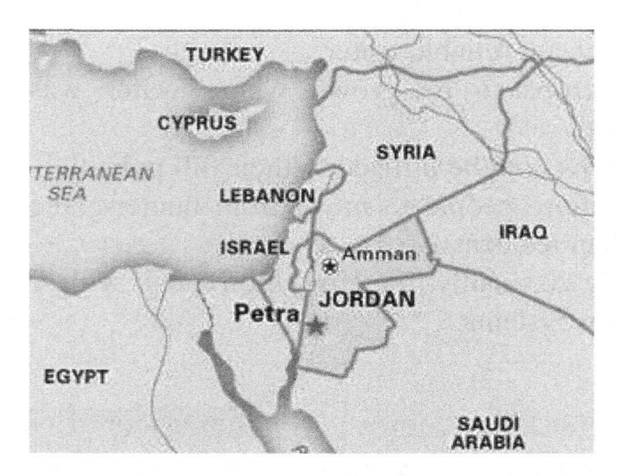

- Built by Nabataeans, an ancient Semitic people dating to 586 BC.
- They created a sophisticated rainwater collection system, and thus, an impressive trading empire.
- Technologies included aqueducts, terraces, dams, cisterns, reservoirs, rainwater/groundwater harvesting, and natural springs harvesting.
- By balancing reservoir water storage with pipelines (seen here), they ensured a constant water supply.
- Technology was used first to benefit the political and economic elites in Nabatean society, eventually filtered down to lower levels of society.

Egypt (Nile River Irrigation, c. 2000 BC)

A "shaduf" used to lift water.

- What can a mural tell us about water policy?
  - Diverse crops were irrigated — e.g., figs, dates.
  - Human-powered technologies were applied to water management.
  - Water use was sometimes regimented, orderly, labor intensive.

Roman Empire — Domestic Water Management

Public latrines — City of Ostia.

Public baths — Pompeii.

- PURPOSE: Romans viewed water management as a means of ensuring public health and sanitation, personal comfort, high quality of life.
- PROCESS: Building and maintaining infrastructure was responsibility of state — *no expense was spared.*

Failures of Ancient Hydraulic Societies

Hohokam culture, c. 1450 AD (Arizona State Univ.).

- Hohokam society — Arizona, No. Mexico (c. 300–1450 AD).
- Innovators in irrigated farming along Gila and Salt Rivers; provided surplus food, allowed regional control.
- C. 1100: Population growth in response to available water supply prompted *stress* at precisely the time protracted drought occurred.
- By 1500 AD, Hohokam disappeared from archeological record — failures in water management contributed to sudden decline on economic and political stability.

For Discussion

- What *positive* lessons for water policy can we take from these experiences?
  That human engineering can make-up for deficiencies in natural water availability as well as control floods.
- What can we learn from any *failures* in these experiences?
  For every engineering fix or other human intervention there can be adverse consequences not foreseen. We therefore need to be prudent and cautious in managing water.
- Are we doing better at managing water today?
  In some respects we have learned from past mistakes. However, in others we continue to build in floodplains, over-rely on structural solutions to drought, and fail to account for the variability of climate.

California as a Hydraulic Society — Balancing Development and Preservation

- California's political and economic history *also* shaped by water: its availability, quality, ecological vulnerability — we've also been called a *hydraulic society* (Hundley, 2002).

- Over 80% of our water is used for agriculture; much of that is "re-used" for cities (together with the remaining 20%).
- Massive public investment — by California and US government — in flood control, irrigation, public supply, hydro-power projects has occurred.
- A perennial debate — how do we develop water resources to *efficiently* maintain agriculture and cities while *preserving* scenic rivers, inland lakes, and threatened flora and fauna and *equitably protecting* everyone's water needs?

California as Hydraulic Society

- World's 5th largest economy was built on harnessing water to economic growth.
    - Traditionally, powerful interest groups exercise control over water and infrastructure, as well as law and policy.
    - GOALS? Security, stability, prosperity, environmental quality.

California as Hydraulic Society — a More Provocative View

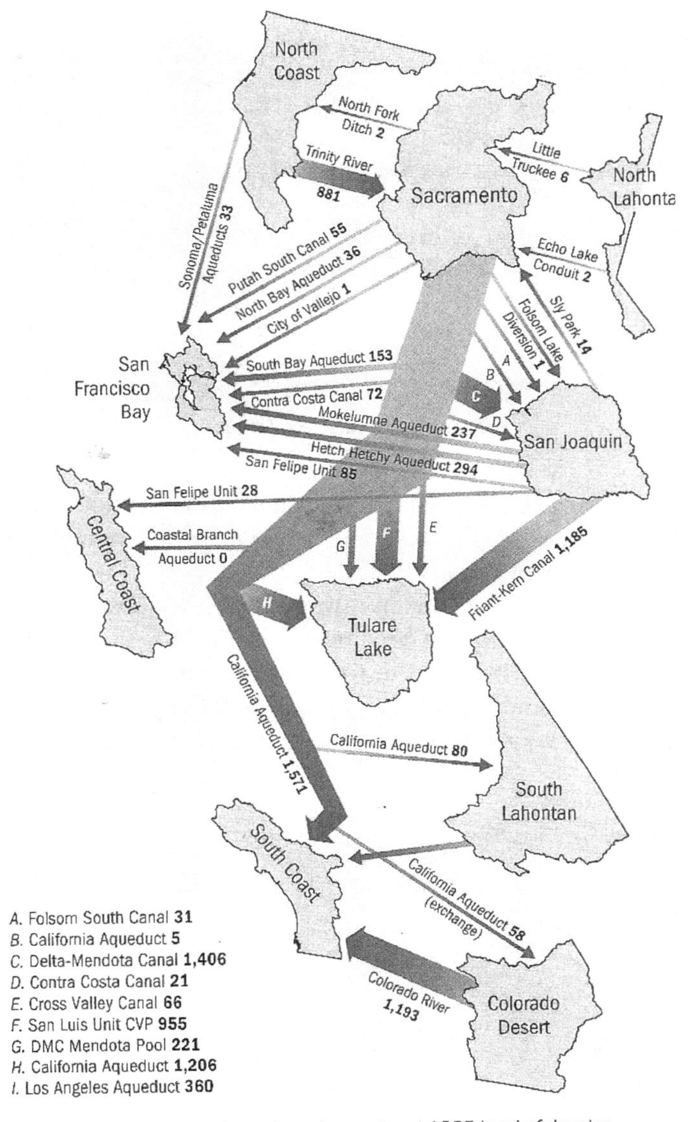

Map 11. Regional water imports and exports, at 1995 level of development (thousands of acre-feet per year). (Redrawn from California Department of Water Resources 1998.)

How Did the First Californian's Manage Water?

Chumash village rendering

- Prior to European conquest, California tribes understood seasonality, relied on coastal resources, had a "symbiotic" relationship with nature — a "water-sensitive" ethic.
- Water claims forcibly taken away after conquest (c. 1770) — worsened when California became a state (1850).
- Today numerous tribal suits to force settlement of Indian water rights claims under Winters Doctrine (1908) when US Supreme Court ruled *tribes have "reserved rights"*.

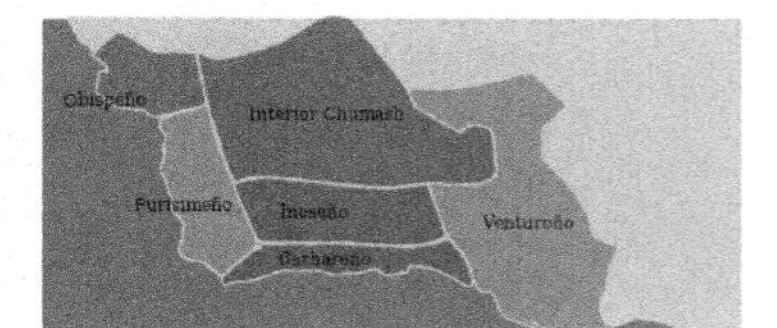

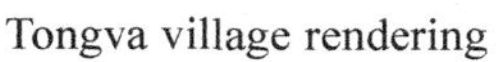

Tongva village rendering

Massive investments transformed our state into a large exporter of food and fiber via massive engineering projects.

Crop irrigation in Central Valley and California Aqueduct — Central Valley project.

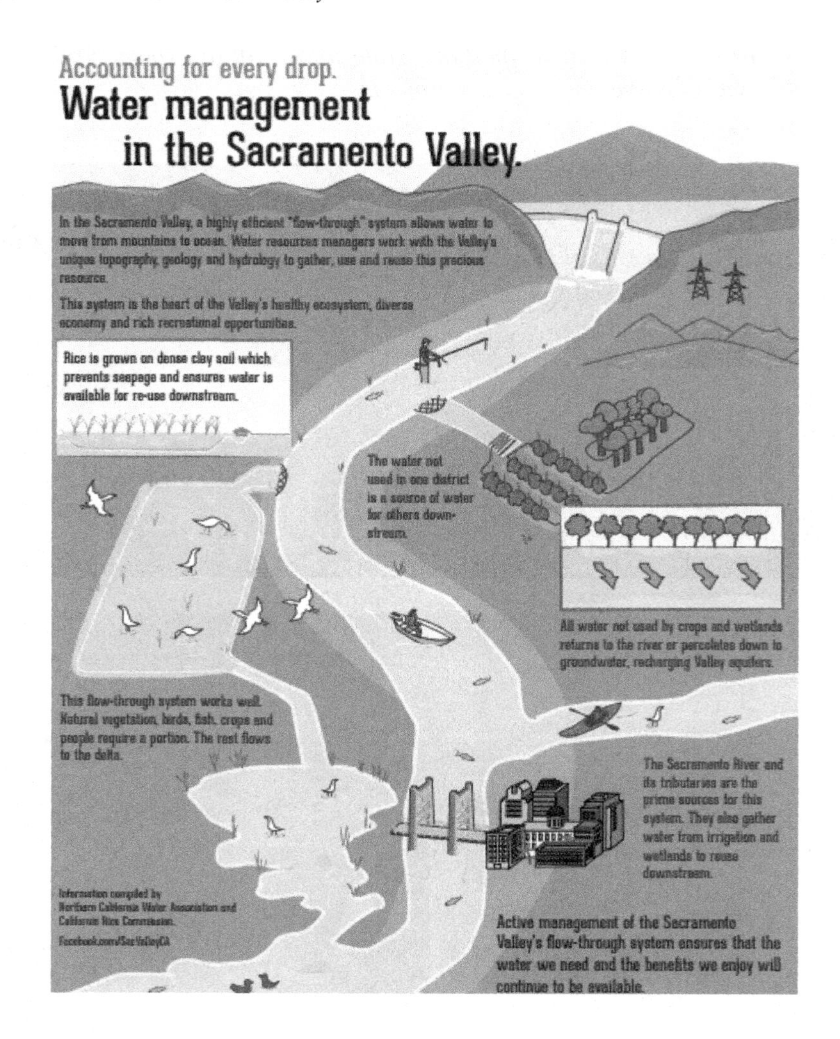
Accounting for every drop.
Water management in the Sacramento Valley.
In the Sacramento Valley, a highly efficient "flow-through" system allows water to move from mountains to ocean. Water resources managers work with the Valley's unique topography, geology and hydrology to gather, use and reuse this precious resource.
This system is the heart of the Valley's healthy ecosystem, diverse economy and rich recreational opportunities.
Rice is grown on dense clay soil which prevents seepage and ensures water is available for re-use downstream.
The water not used in one district is a source of water for others down-stream.
All water not used by crops and wetlands returns to the river or percolates down to groundwater, recharging Valley aquifers.
This flow-through system works well. Natural vegetation, birds, fish, crops and people require a portion. The rest flows to the delta.
The Sacramento River and its tributaries are the prime sources for this system. They also gather water from irrigation and wetlands to reuse downstream.
Information compiled by Northern California Water Association and California Rice Commission.
Facebook.com/SacValleyCA
Active management of the Sacramento Valley's flow-through system ensures that the water we need and the benefits we enjoy will continue to be available.

Providing Urban Supply — San Francisco

Hetch Hetchy River — c. 1900.

- In early 1900s, San Francisco was state's largest city but was fresh-water-limited; sought to expand water supply — QUESTION: should preserving wilderness (Yosemite Park) or damming rivers take priority?
- Major protagonists — John Muir and Gifford Pinchot (dispute gave birth to Sierra Club).

Hetch Hetchy Valley — 2020

- *O'Shaughnessy Dam* (completed 1913) — supported by Gifford Pinchot, Teddy Roosevelt.
- TODAY: Some favor removing the dam — others want to leave it as is.
- 2012: Bay area voters decided on Proposition F — whether to demolish the dam and identify replacement water and power sources — was defeated 77%–23%.
- 2018: Supporters of dam removal (Friends of Hetch Hetchy) took case to California Supreme Court — court sided with state that permission to build the dam was *a federal, not state issue*.

Hetch Hetchy — Urban Water Diversion

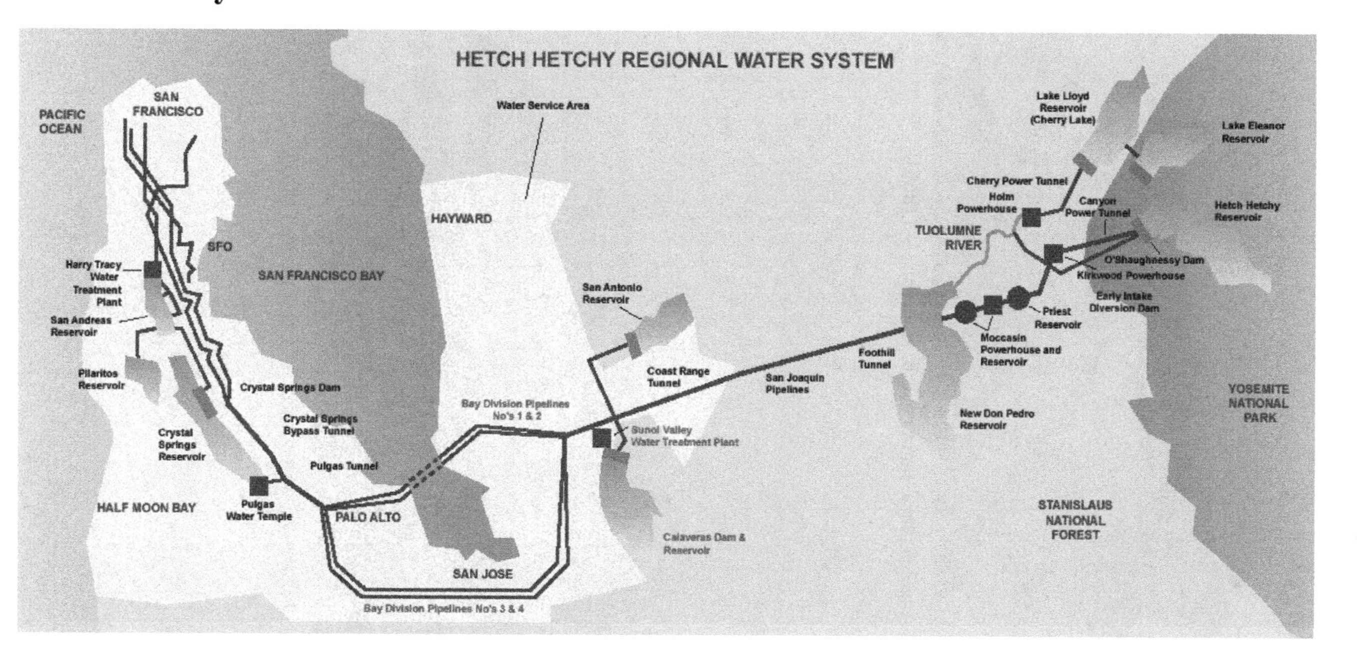

… And Los Angeles

Los Angeles plaza, main church, reservoir — (1869).

Zanja Madre and water wheel — Los Angeles (1863).

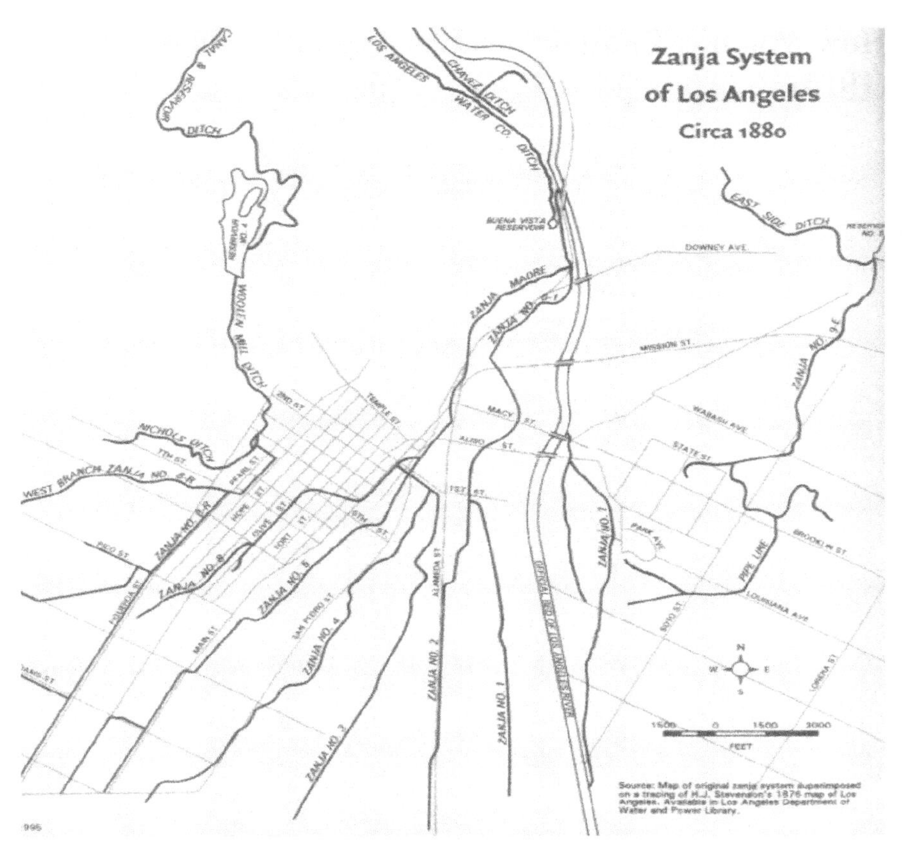

Zanjas, literally "ditches", diverted water from the Los Angeles River to provide supplies for homes, businesses, orchards and local farms. Until the early 20th Century, these Zanjas, together with some groundwater, were the city's only source of supply.

How Much is Enough? — Los Angeles Aqueduct (1913)

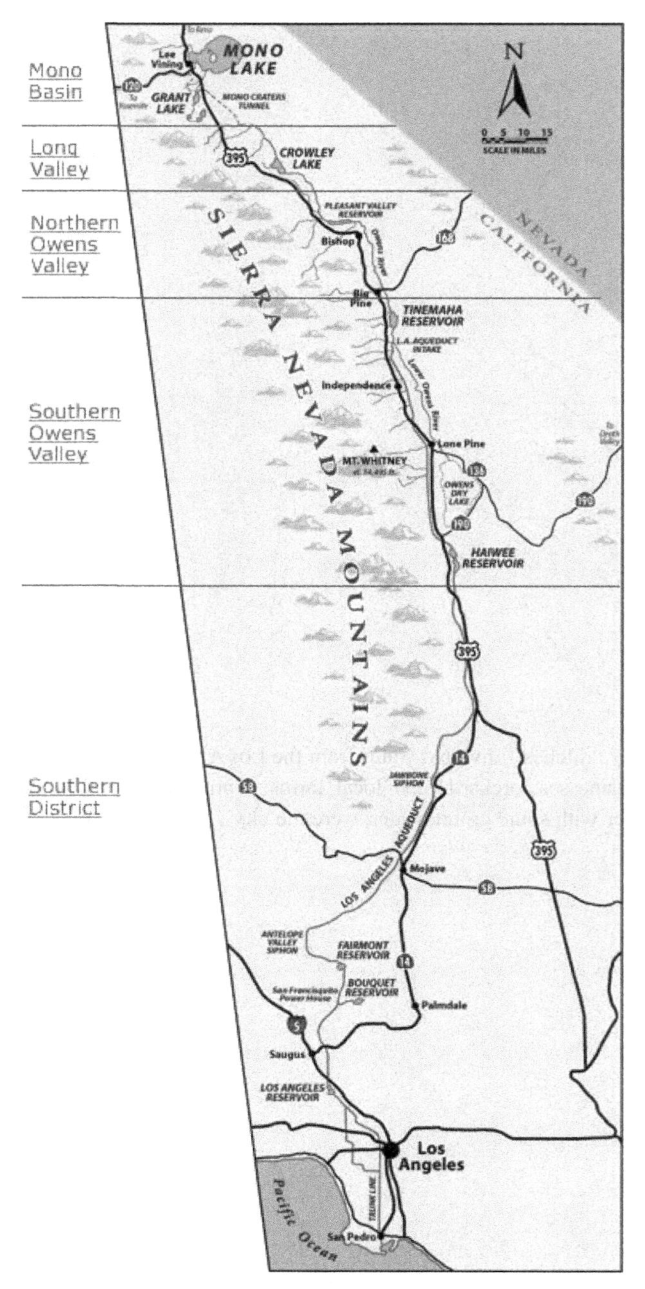

- By early 20th century, LA's population doubled every 10 years — Los Angeles River no longer an adequate water source.
- Civic leaders sought reliable sources whose rights they could easily obtain — Owens Valley was surveyed, land and water rights secretly acquired. *Problems this created*?
    - *Whose* economic development aspirations should prevail? (agriculture effectively came to an end in Owens Valley).
    - *How* should decisions over water allocation be made — inclusively or exclusively? (land and water acquisitions were secret).
    - *How* much water was needed for Los Angeles vs. the environment, and Owens Valley residents? (still debated!).

JP Lippincott, Fred Eaton, William Mulholland — c. 1910.

Building the Owens Valley Aqueduct — 1908–1913

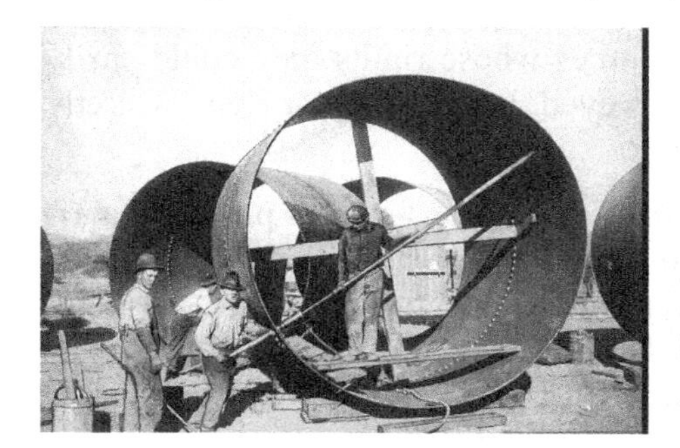

November 1913 — Opening of aqueduct: "there it is, take it" — W. Mulholland.

Terminus of Aqueduct near Sylmar.

There is an enormous amount we can learn from the management and governance of water throughout history. The most significant lessons are three-fold:

1. Power to make decisions is generally unequal - with economic development interests and established water bureaucracies often directing decisions. Nonetheless, accommodation among interests favoring and opposing innovations typically occurs due to the need for consensus.

2. The process of decision-making has often been technocratically-driven, favoring certain types of engineering or other technical expertise and thus, a relatively narrow set of voices.
3. *Purpose* is the most elusive component of water governance. Protagonists throughout history often disagree on the objectives of water policies, and on water supply projects of various types — from dams to desalination plants.

St. Francis Dam — Storage for the Los Angeles Aqueduct

May 1928

Local Reaction to Los Angeles Aqueduct — 1920s

Peace Put Under Way in War Over City's Aqueduct

Ranchers of Owens Valley "Sitting Tight"

CAMP AT AQUEDUCT GATE IS CENTER OF FAMILY LIFE

Some Women Cook for Watchers While Others Care for Tots; Girls Form Orchestra

Los Angeles Daily Times, November 20, 1924. After four days, *Times* photographers had been to the occupation site and returned home with their film. Photographic layouts and stories described the picnic at Alabama Gates, portraying the encampment of "350 ranchers" as a center of family life, entertainment, and local solidarity. (Courtesy of the *Los Angeles Times*.)

- Regional resentments arose over aqueduct's impacts on local economy; sense of "betrayal".
- Lessons? Making water policy secretly and without consultation with all affected leads to environmentally unjust outcomes.
- Some residual resentments remain.

Harbors and Ports — Toward a Pacific Empire

- In 19th century, political fights erupted over harbors in SoCal — some wanted it in Santa Monica, others San Pedro, still others chose San Diego.
- 1897 — Debate settled by Congress-appointed "Board of Engineers" — recommended San Pedro be expanded *and* offered to pay for expansion.
- 1909 — LA annexed San Pedro and Wilmington to assure political control of harbor. It's now the largest port complex in Western Hemisphere.

Lumber docks — San Pedro — 1887.

Los Angeles Harbor today.

Legacies of California's Hydraulic Society

Alkali dust storm over Owens Lake.

FROM: "L.A. took their water and land a century ago. Now the Owens Valley is fighting back," *Los Angeles Times*, July 13, 2017.

- Diversion of Owens Valley depleted Owens and Mono Lakes; created toxic dust plumes and ecological degradation (selenium, cadmium, arsenic).
- 1994 settlement with EPA, LADWP, Mono and Inyo counties — required restoration of Owens River; restocking of bass, bluegill, trout.
- 2013 *Stream Restoration Agreement* required Los Angeles to restore stream-flows to historic levels. Public protest was key to policy change.
- Since 2014 LA and Owens Valley work together to *monitor Owens and Mono Lakes*, determine when, and how much water can be released.
- *However*, regional income remains dependent on tourism, NOT water supply for agriculture.

Other Legacies — The Gold Rush and Environmental Devastation

Mid-19th Century gold miners.

- Erosion and water depletion were extensively generated by hydraulic mining.
- Explains California's adoption of laws to promote "regulated withdrawal" + "beneficial use" (beginning in 1880s).

The Public Trust Doctrine — Balancing Development and Environment

- 1928 California constitution requires "highest beneficial use" of water — a *response to the Owens Valley controversy.*
  - "Watershed of origin" rights take precedence over rights of "exporters" (thus, affording protection against impacts of diversion).
  - Rights to water are determined *in part by ownership of land* adjacent to a water body, but also by *timing* of use — some rights holders are more "senior" and thus, entitled to more water.
  - 1994: *Mono Lake case* — California supreme court fully instituted *Public Trust Doctrine* — state has a duty to protect "the people's common heritage of streams, lakes, marshlands, and tidelands . . . ."

California as Hydraulic Society — Continuing Challenges

- PROCESS — there is no real "arbiter" among various water policy interests — e.g., environment vs. economics. Thus policies tend to be fought out locally, between groups with different interests.
- PURPOSE — priorities still diverge; e.g., development vs. preservation; agriculture vs. cities (e.g., Hetch Hetchy, Owens Valley) — can views be reconciled?
- POWER — water policy tends to be interest-driven. In US and California, many water policies are driven by groups with inordinate influence (i.e., money, status, political advantage).
    - *As in antiquity*, water remains a key to regional development; if political leaders fail to satisfy wishes of constituents, dissatisfaction may result — especially during times of drought or other crisis.
    - Seen today in *many* proposals for protecting the state's water supply — e.g., Governor Newsom's Water Resilience Portfolio.

"Water is central to nearly everything we value in California. Healthy communities, economies, farms, ecosystems and cultural traditions depend on steady supplies of safe and affordable water".

Water Resilience Portfolio — Executive Summary, January 2020. http://waterresilience.ca.gov/wp-content/uploads/2020/01/California-Water-Resilience-Portfolio-2019-Final2.pdf

Final questions and observations?

Section 3

US Policies in a Global Context — Law, Practice, and Institutions

United States — Average Annual Precipitation

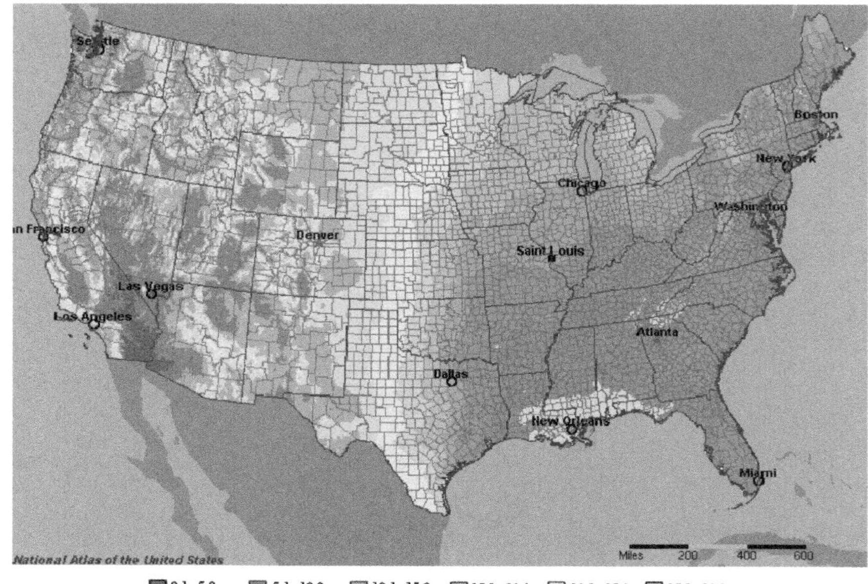

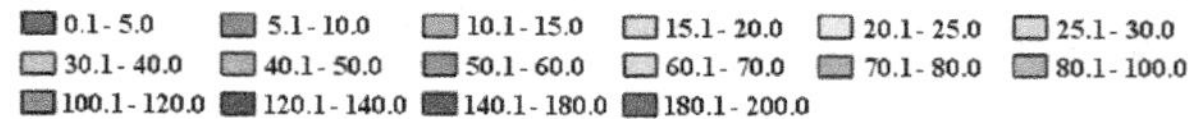

The 100th Meridian — A Geographic and Hydrological Dividing Line

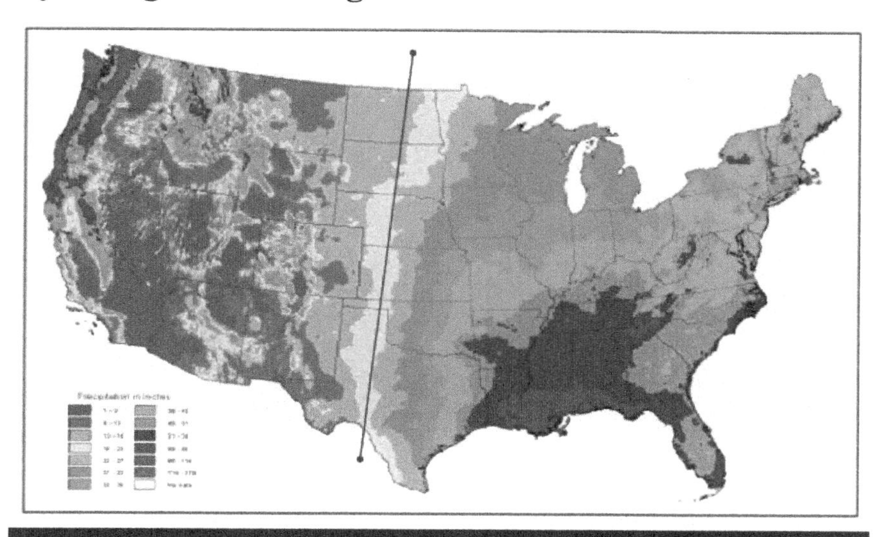

The approximate location of the 100th Meridian... the traditional dividing line between the humid east and the arid west

Water Law — How Water Supply is Allocated

- *Riparian doctrine (East)*: property owners are entitled to "reasonable use" subject to rights of downstream users — *but not a specific amount of water.*
- *Prior appropriation (West)*: "first in time, first in right". The *first* user of a water body has priority over subsequent users — *and are entitled to a specified amount.*
  - California has modified prior appropriation by formally distinguishing "senior" from "junior" rights holders.
  - Users must use water "beneficially" — take what you need, but prove you *need* what you take, and *don't inflict harm on environment or other users.*
  - State can "pre-empt" individual rights for public needs (e.g., drought).
- *Hydrology has partly dictated law and policy!*

Riparian Rights — an Illustration

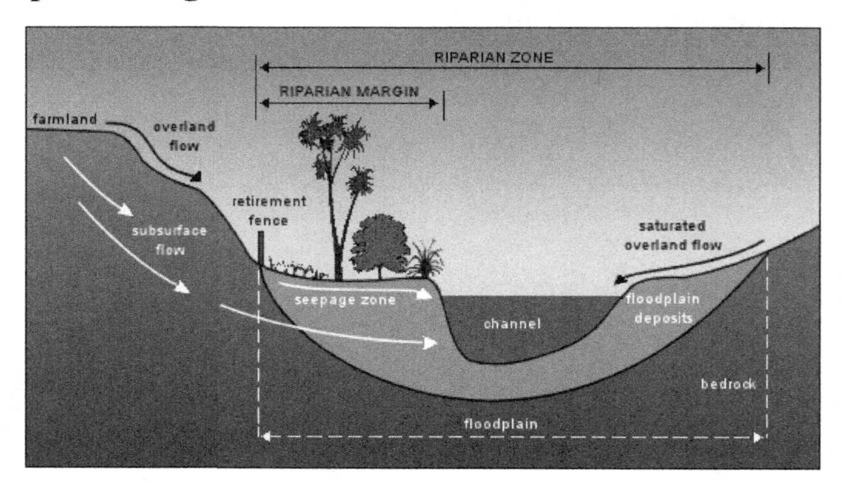

- *Riparian rights are property rights* — they inhere in a riparian parcel of land.
- A parcel of land must border a natural water body to be identified as riparian.
- If you own a riparian parcel of land you control:
  - The *upland.*
  - Any structures in the riparian zone.
  - The *bottomland* offshore from your lot.
  - The *aquatic vegetation* growing on your bottomland.
  - Rights to *groundwater* flowing into the "seepage zone".

Prior Appropriation Rights — A California Illustration

Prior Appropriation: an example
"First in time, first in right"

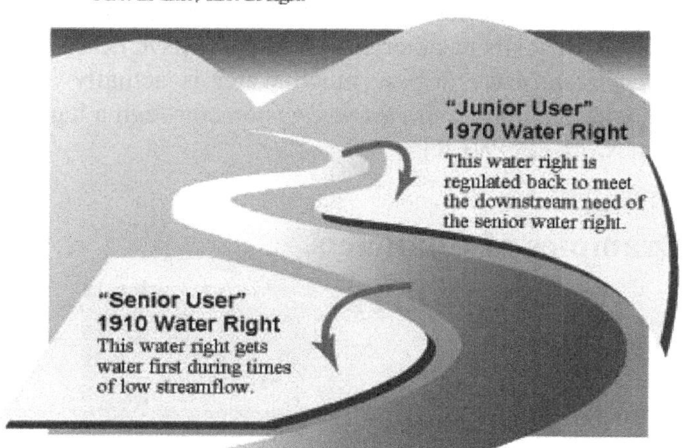

An example of prior appropriation at work
Prior appropriation ensures that the first water user to obtain water rights has first access to water in times of shortage. If a "downstream" landowner has the earlier priority date (they initiated their water right in 1910) the "upstream" landowner may have to let the water pass unused to meet the needs of the senior, downstream water right holder.

- *Appropriation rights are property rights* — based on intent to withdraw. Users:
  - Must exercise right to use — if you fail to use the water, you can lose your right to it.
  - May sell or lease rights to the water you don't use (called — water marketing).
  - Must use entitlement beneficially — i.e., not wastefully, and with respect for environmental and navigation needs of region (i.e., public trust).

Moving Beyond Conventional Law — Compacts

- *Neither riparian nor appropriation doctrines* ensure that water be fairly shared — every state (and country) sharing a river basin has slightly different laws, practices, and policies.

- *Compacts* are agreements among different countries or states within-a-country to:
  - Provide *integrated governance* of a watershed or basin — to allocate supply, regulate demand, and protect quality (PURPOSE).
  - Equitably manage how water is shared by *consulting* with government officials & other stakeholders (PROCESS).
  - *Independently verify* how much water is actually available for use, and to *enforce allocation decisions* through a legally-binding framework (POWER).

Some Examples of Compacts

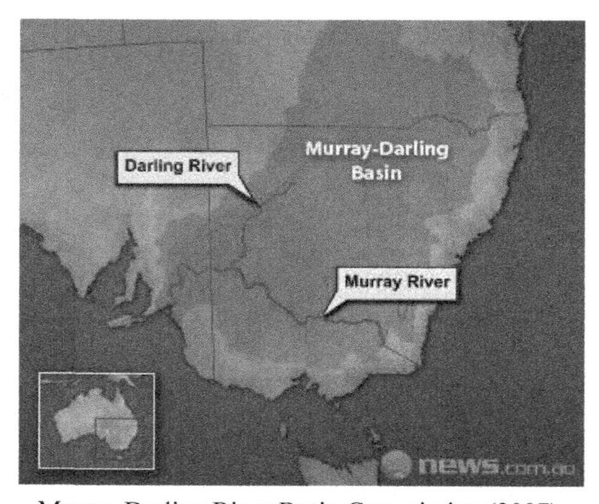

Murray Darling River Basin Commission (2007).

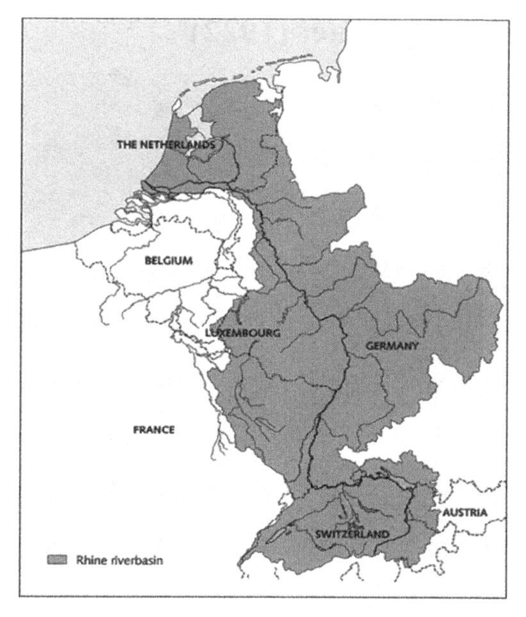

International convention on protection of the Rhine (1999).

Delaware River Basin Commission (1965).

Colorado River Compact (1922)

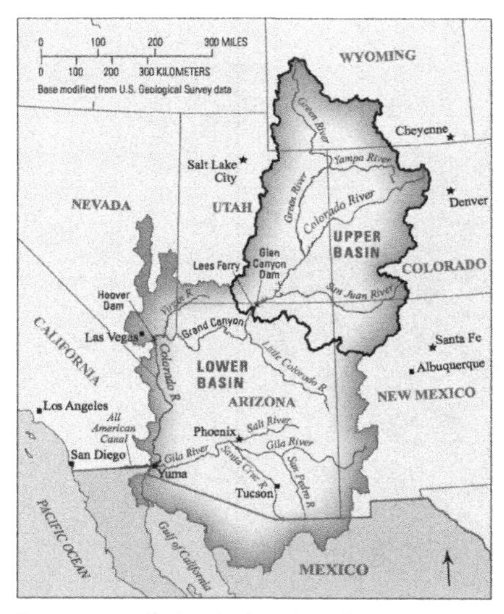

Base map of physical and political features of Colorado basin (USGS).

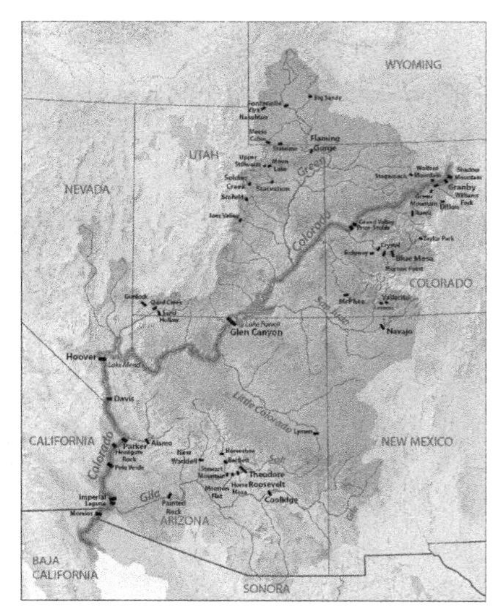

Dams in Colorado basin (2017).

Colorado River Compact — Overview

- In 1922, seven basin states and the federal government (Arizona,* California, Colorado, Nevada, New Mexico, Utah, Wyoming) negotiated agreement to allocate flow:
  - *Between upper (CO, UT, WY) and lower basins* (AZ, CA, NV) — (NM in both).
  - Among states within entire basin — based on prior uses and estimated future needs.
- Negotiators assumed an annual flow of 14–17 million acre-feet (MAF) of water/year — with an average at 16.5 MAF/year.
- If true, this would make it possible for *upper and lower basins* to be allocated 7.5 MAF/year with the remaining 1.5 MAF available for *Mexico*.
- Negotiators got it wrong *climatically* and *politically*.

* Initially refused to sign compact (until 1944).

Politics vs. Science — Climate Variability on the Colorado River

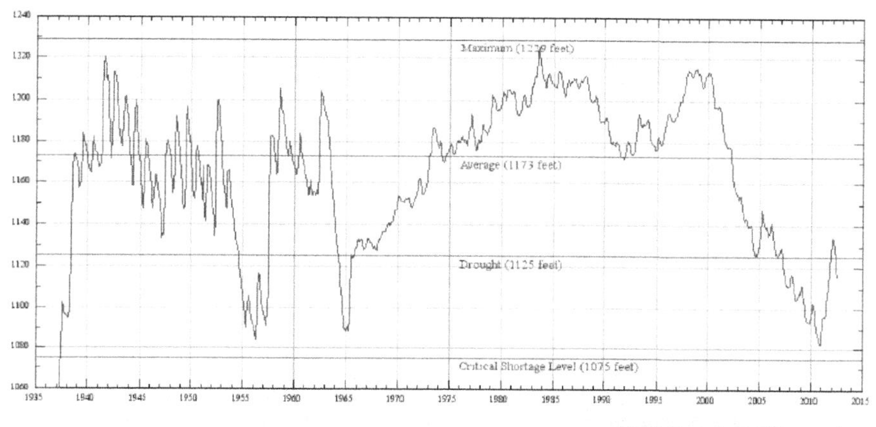

Lake Mead water levels: 1935 – present (US Bur Rec)

- ***<u>Colorado River flow is less than negotiated</u>*: in most years between 11-13 MAF.**
- ***<u>Calculations based on contemporary measurements</u>*: 1920s were high flow years – *dendrochronology* proved flows were *exceptional* due *to* recurrent drought, climate variability.**

Politically — California's Thirst wasn't Fully Appreciated

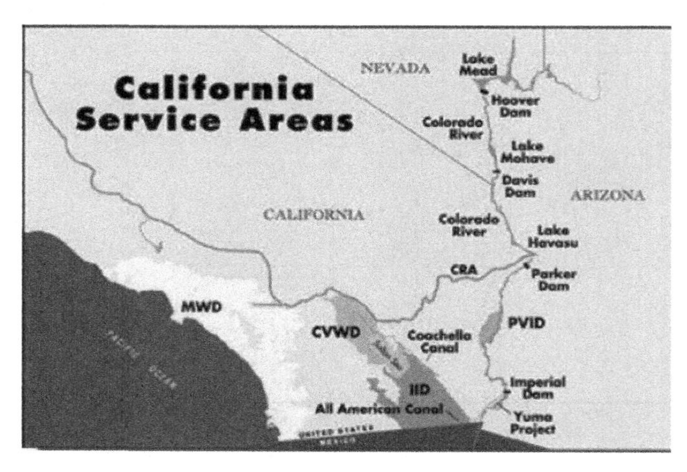

KEY:
MWD – Metropolitan Water District (LA, Orange, San Diego, portions of Ventura, Riverside & San Bernardino Counties)
CVWD – Coachella Valley Water District
IID – Imperial Irrigation District
PVID – Palo Verde Irrigation District
CRA – Colorado River Aqueduct

CALIFORNIA IS A MAJOR USER OF COLORADO RIVER WATER

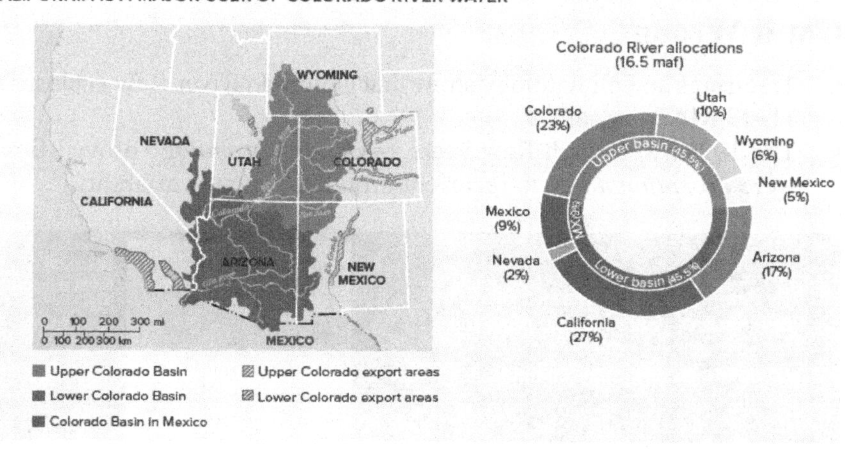

WATER USE HAS BEEN OUTSTRIPPING SUPPLY IN THE BASIN

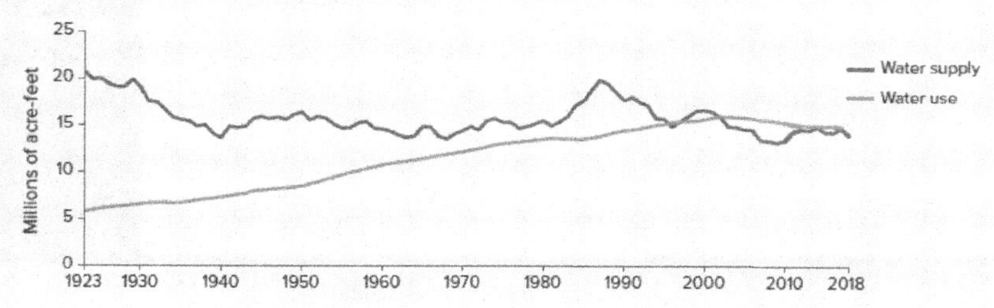

SOURCE: Adapted from US Bureau of Reclamation (USBR), *Colorado River Basin Water Supply and Demand Study* (2012), with updated historical supply and use data from USBR.

NOTES: The figure shows water use and supply as 10-year running averages. Supply estimates for 2016–18 and use estimates for 2017 are provisional.

Science vs. Politics — How Negotiators Got it Wrong

- Tree-rings and climatology show that Colorado River flow is closer to 11–13 MAF annually .
- The 1920s were high flow years — *dendrochronology* proved flows were *exceptional* due to recurrent drought, climate variability.

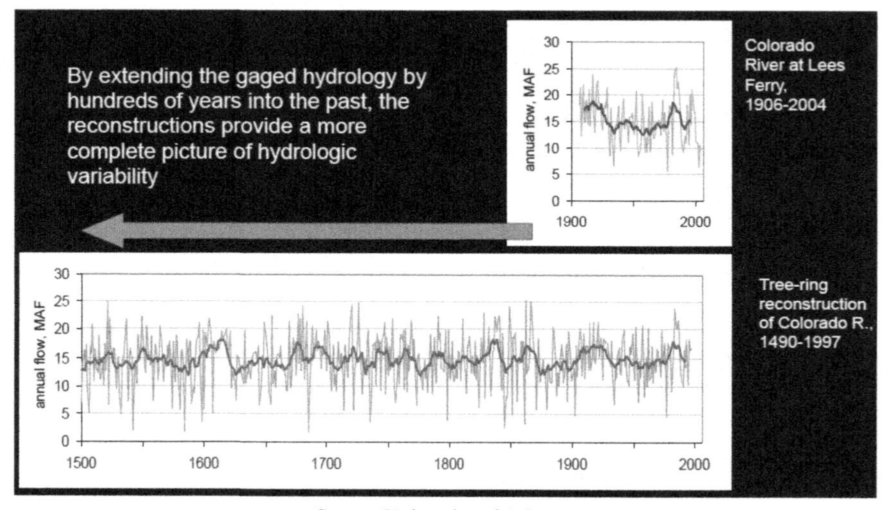

Source: University of Arizona.

Remaining Challenges in Getting the Science and Politics Right

ENVIRONMENT

'The pie keeps shrinking': Lake Mead's low level will trigger water cutbacks for Arizona, Nevada

Ian James Arizona Republic
Published 7:30 a.m. MT Aug. 15, 2020

Arizona's state-approved plan for managing the cutbacks involves delivering "mitigation" water to help lessen the blow for some farmers and other entities, while making compensation payments to those that contribute water. Officials have said the state will spend approximately $59 million on the plan, and the Central Arizona Water Conservation District has pledged about $65 million.

Much of the money will go toward paying for water contributed by the Colorado River Indian Tribes and the Gila River Indian Community.

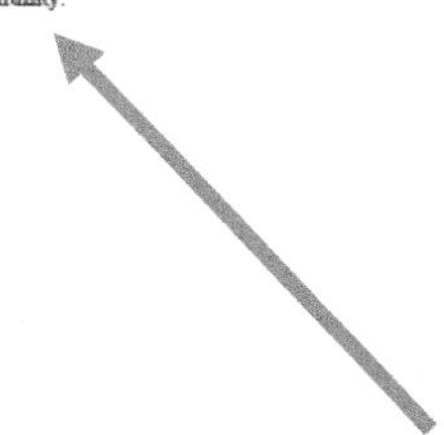

Arizona, Nevada and Mexico will again receive less water from the Colorado River next year under a set of agreements intended to help boost the level of Lake Mead, which now stands at just 40% of its full capacity.

The federal Bureau of Reclamation released projections on Friday showing that Lake Mead, the nation's largest reservoir, will be at levels next year that continue to trigger moderate cutbacks in the two U.S. states and Mexico.

Those reductions could eventually increase in the next few years if Lake Mead drops further. The estimates show the reservoir near Las Vegas will likely begin 2021 about 10 feet above a level that would trigger larger cuts.

For more stories that matter, subscribe to azcentral.com.

The Colorado River, which supplies cities and farmland from Wyoming to Southern California, has long been chronically overused and has dwindled during two decades of mostly dry years. The drought has been worsened by higher temperatures unleashed by climate change.

The reductions to Arizona and Nevada are part of a 2019 deal called the Lower Basin Drought Contingency Plan, which also calls for California to face cuts if reservoir levels continue to fall. Under a separate accord, Mexico will contribute next year by leaving some of its allotted water in Lake Mead.

The federal government's forecast also estimated the levels of Lake Powell along the Arizona-Utah border. That reservoir, the country's second largest, now sits 50% full.

The projections show that Lake Mead, while remaining above an official shortage level, will for a second year be within a zone called "Tier Zero," which brings the initial cuts in Arizona and Nevada.

Drought has reduced the amount of water on the Colorado River, draining storage in Lake Mead. Mark Henle/The Republic

If the states that rely on the river hadn't reached the shortage-sharing agreements, the federal government's data shows that Lake Mead would have dropped significantly lower, said Sarah Porter, director of Arizona State University's Kyl Center for Water Policy.

For Discussion

- What should negotiators or the Colorado River Compact have done to avoid these problems?
- What obstacles to better use of science did negotiators face, do you think?
- What should we do today to address these problems?

Politics of the River — Arizona vs. California

- *Arizona*: initially refused to ratify compact; claimed it was an "inequitable division of its share of lower Colorado" and denied its right to "full appropriation".
- In 1952, filed a brief with US Supreme Court against California.
- *California response*: Arizona was already diverting the *Gila River*, a Colorado River. tributary — so, it didn't need more water.
- California also claimed its use of Colorado River preceded Arizona's and was, thus, legally "perfected" under *prior appropriation doctrine*.

- In 1934, 12 years after refusing to sign the Colorado River Compact, Arizona's governor, Benjamin Moeur, called up the Arizona National Guard and ordered it to defend the state against California attempts to build a diversion dam (Parker Dam) and to keep any construction from taking place on their river bank.

US Supreme Court Intervenes — 1963

- *Arizona vs. California*: *prior appropriation doctrine* doesn't apply — only US has right to apportion water on a trans-boundary, navigable river.
- In addition:
  - *Arizona* is entitled to fully divert the Gila and a portion of the Colorado River — 2.8 *MAF/year — mostly for Phoenix and Tucson* for cities.
  - California must reduce its allocation to 4.4 MAF/year.
  - Indian reservations are entitled to "reserved rights", as promised under 19th Century treaties — including Quechan, Navajo, Zuni.
  - All lower basin states (AZ, CA, NV) must annually report on their use.
  - In 1979 — Court issued a "supplemental decree" providing that state allocations take priority over efforts since 1922 to re-allocate water rights.

Central Arizona Project — One Result of Arizona vs. California

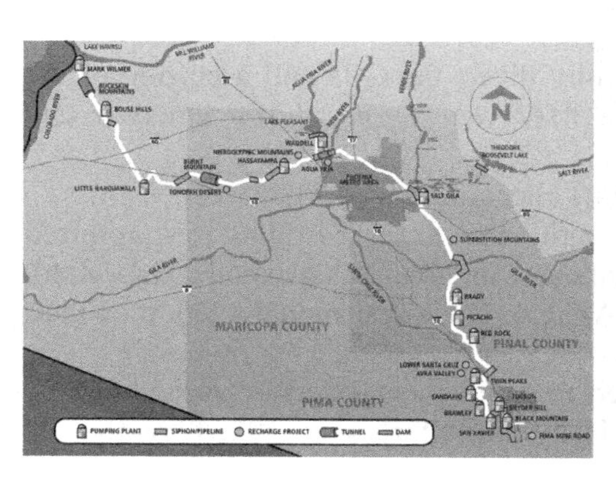

And then there's Las Vegas and Southern Nevada

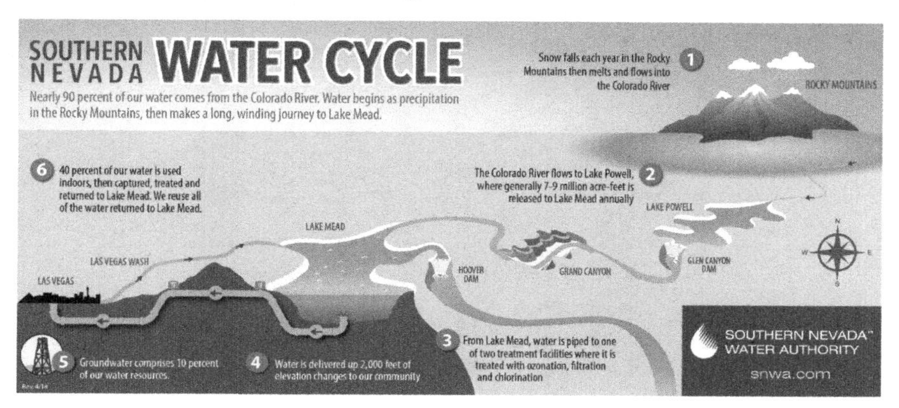

- Under the Colorado River Compact, and as ratified by the US Supreme Court, Nevada has the right to consumptively use 300,000 acre-feet of water per year. An acre-foot is the amount of water that can cover an acre to a height of one foot and is equivalent to 325,851 gallons.
- Consumptive use is defined as water withdrawals (or diversions) minus any water that is returned to the Colorado River. These returns are also known as return flow credits.

More Conflict — Mexico vs. U.S.

- *Colorado River is a bi-national resource*:
  - Mexico threatened to go to *International Court of Arbitration to ensure its fair share*.
  - *Mexico–US treaty* (1944) requires that US provide 1.5 MAF/year to Baja and Sonora.
  - To ensure goal is met, US built a desalination plant (Yuma, AZ — 1992).
  - Water deliveries are reliable, but problematic — due to drought, low streamflow.

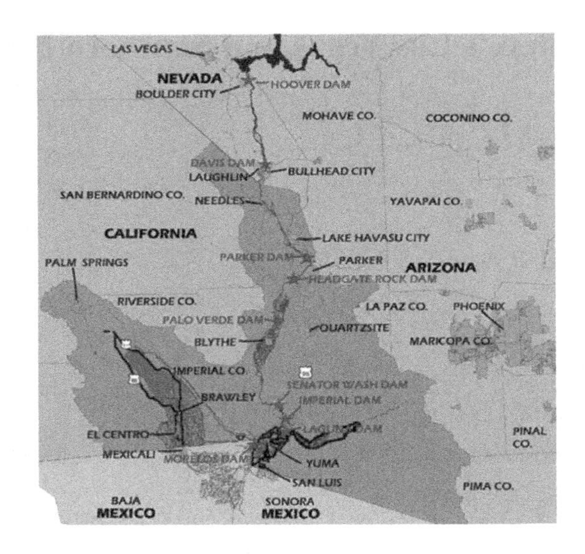
LAS VEGAS
NEVADA
BOULDER CITY
HOOVER DAM
MOHAVE CO.
COCONINO CO.
DAVIS DAM
LAUGHLIN
BULLHEAD CITY
SAN BERNARDINO CO.
NEEDLES
YAVAPAI CO.
CALIFORNIA
LAKE HAVASU CITY
PALM SPRINGS
PARKER DAM
PARKER
ARIZONA
HEADGATE ROCK DAM
RIVERSIDE CO.
LA PAZ CO.
PHOENIX
PALO VERDE DAM
QUARTZSITE
BLYTHE
MARICOPA CO.
IMPERIAL CO.
SENATOR WASH DAM
BRAWLEY
IMPERIAL DAM
EL CENTRO
LAGUNA DAM
PINAL CO.
MEXICALI
MORELOS DAM
YUMA
SAN LUIS
PIMA CO.
BAJA
MEXICO
SONORA
MEXICO

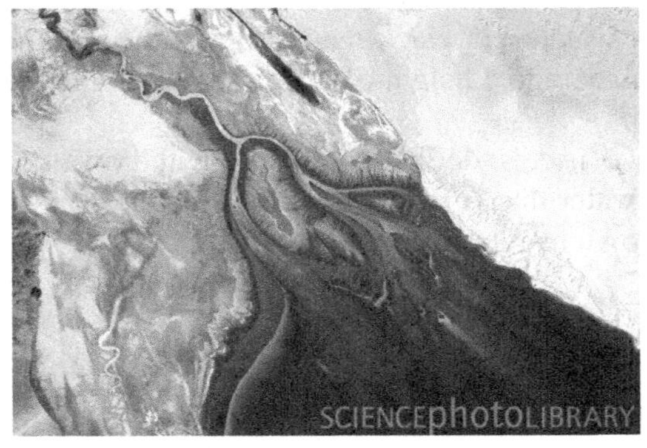
SCIENCEphotoLIBRARY

How Bad can Bi-National Water Conflicts Become?

WORLD & NATION

Mexican water wars: Dam seized, troops deployed, at least one killed in protests about sharing with U.S.

Troops guard a dam Thursday in Camargo, Mexico. (Christian Chavez / Associated Press)

By PATRICK J. MCDONNELL | MEXICO CITY BUREAU CHIEF
SEP. 11, 2020 | 7:08 PM

MEXICO CITY — Mexico's water wars have turned deadly.

A long-simmering dispute about shared water rights between Mexico and the United States has erupted into open clashes pitting Mexican National Guard troops against farmers, ranchers and others who seized a dam in northern Chihuahua state.

A 35-year-old mother of three was shot dead and her husband seriously wounded in what the Chihuahua state government labeled unprovoked National Guard gunfire.

"Minute 323" — 1944 Treaty Addition: Significance?

Dignitaries gather in March 2014, on Morelos Dam in Algodones, Mexico — celebrating agreement on "Minute 323 to allows a "pulse flows" of Colorado River to reach the river's delta in Baja Mexico for first time since 1960s.

- Signed in *September 2017*, in Santa Fe, New Mexico — first time two nations agreed to allocate the river on behalf of *environment*.
- US will invest in water conservation projects in Mexico — e.g., plugging leaks in irrigation canals and helping farmers implement water-efficient technology.
- Mexico can store some water in the US In turn, US will receive a portion of Colorado River water Mexico isn't using.
- Through 2026 — 210,000 acre-feet of water, will be allowed to flow through the lower reaches of the Colorado upstream from Sea of Cortez.

For Discussion

- What kind of a future would you envision for the region if these problems continue?
- What are some the pressures you could envision for the future of the region?

Conclusions — An Uncertain Future

- *Arizona*: has won right to use its "full allocation" of Colorado River.
- *California*: *has had to seek other alternatives* — e.g., diversion from Bay-Delta and Northern California (e.g., *State Water Project*).
- *Nevada*: can divert some *300,000 acre-feet/year* — mostly for Las Vegas — but it's not enough for Las Vegas' projected growth — plans to divert groundwater from Northern Nevada are underway.
- *Upper basin states*: are increasing their use of assigned allocations due to growing energy demands (WY, CO, UT).
- River is over-allocated — every party is using its full share; climate change will worsen problem. Solutions? Greater municipal conservation, re-use, water banking, inland desalination (can possibly "acquire" another 4.2 MAF).*

*American Rivers, 2014; Bureau of Reclamation, 2012.

Section 4

Who Controls Freshwater? Allocation and Uses

Who "Owns" Your Water?

- Those of us living in cities have no real "rights" to water like the landowners we discussed last week — we receive our water from providers whose services we pay for.
- Who are they?
  - *Public utilities*: special government agencies that use taxes and fees to provide and treat water. *Where*: US, Canada, Australia.
  - *Private companies*: stockholder-owned, for-profit entities that charge fees to supply/treat water. Fees cover costs and provide return-on-investment. *Where*: Much of Europe, UK, Latin America, E. Asia.
- All countries have a mix of public/private providers.

Water Provision — Southern California and Orange County

Key:
MWD — Metropolitan Water District (serves LA, Orange, San Diego, portions of Ventura and Riverside counties.

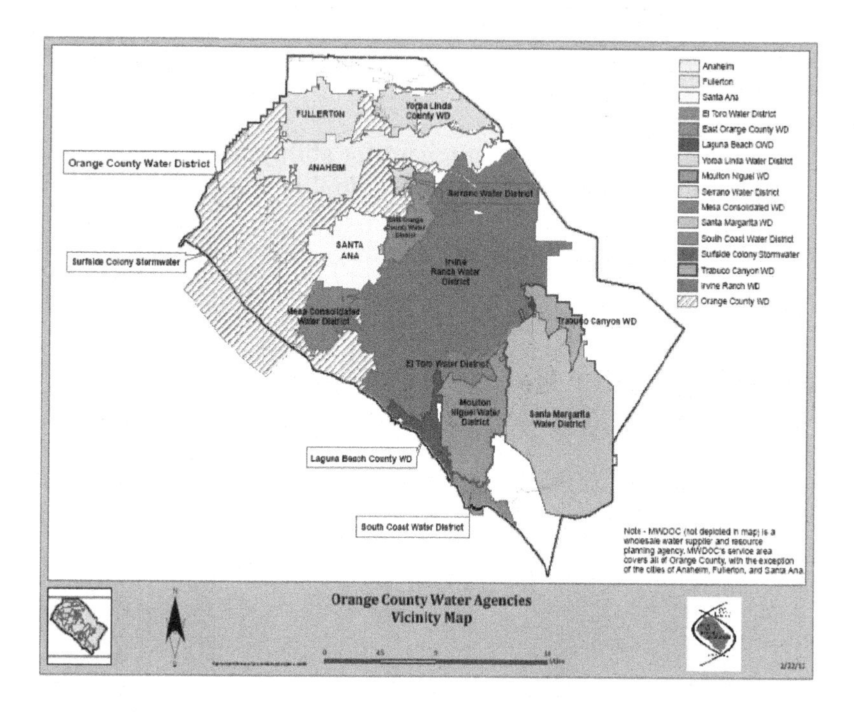

Comparing Public and Private Water Service

- Public agencies are *monopolies* — can keep prices low/economies of scale high.
- Operate on *narrow* earnings margins — enough to recover costs of service.
- Critics say they:
    - Don't encourage conservation for fear of losing revenue.
    - Have high unaccounted water losses (>20% in older cities) — leaks often go unrepaired.
    - Have elected governing boards — they want to keep rates low and are slow to innovate.
- Private companies are also monopolies — service more reliable, but expensive.

- Operate on *high* earnings margins to provide stockholders with return-on-investment.
- Critics say they:
  - Maximize profits without regard for ability-topay.
  - Tend to operate where profit potential is high, current service is poor.
  - Have non-elected corporate boards — less responsive and *transparent* to public.

Comparing Water Rates — US

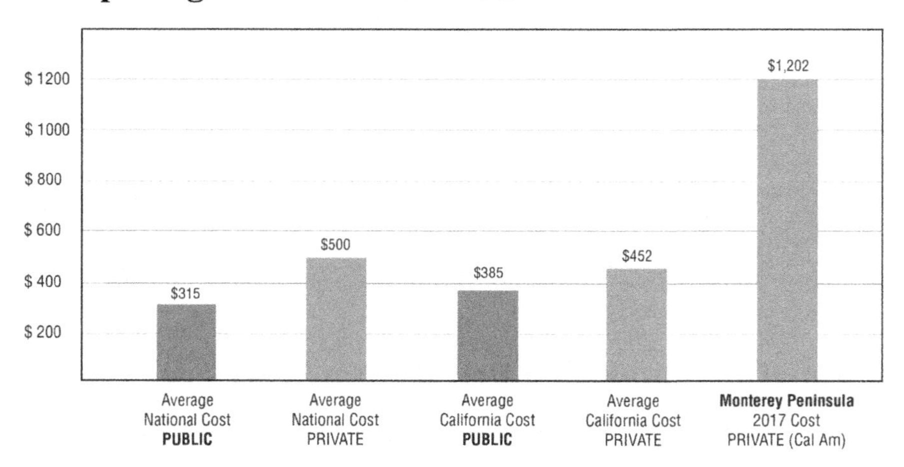

Comparing Water Rates — Internationally

Country	Cost per cubic meter (in US dollars)	Public or private providers predominate?
France	1.23	Private — 3 dominant vendors
United Kingdom	1.18	Private — 35 vendors
United States	0.40	Public — thousands of providers
Canada	0.51	Public — 100 + providers

Source: UN World Water Development report.

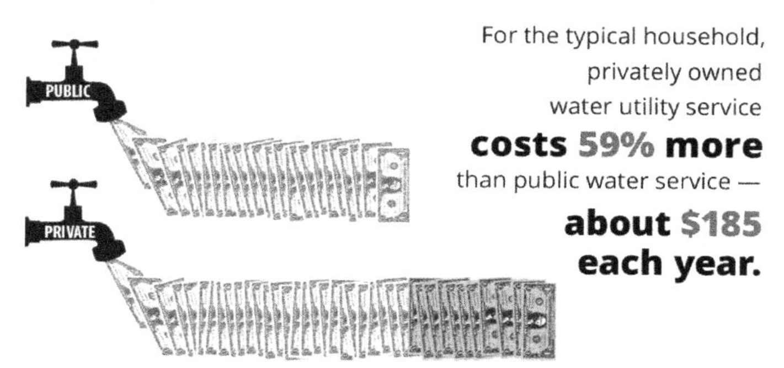

Source: Food and Water Watch Fund — http://www.foodandwaterwatch.org/.

US Gradually Moving Toward Greater Privatization — Why?

- Growing cost of regulatory compliance.
- Urban sprawl — need to expand distribution systems.
- Costs of removing contaminants-of-concern.
- Federal budget cuts for water treatment and supply.

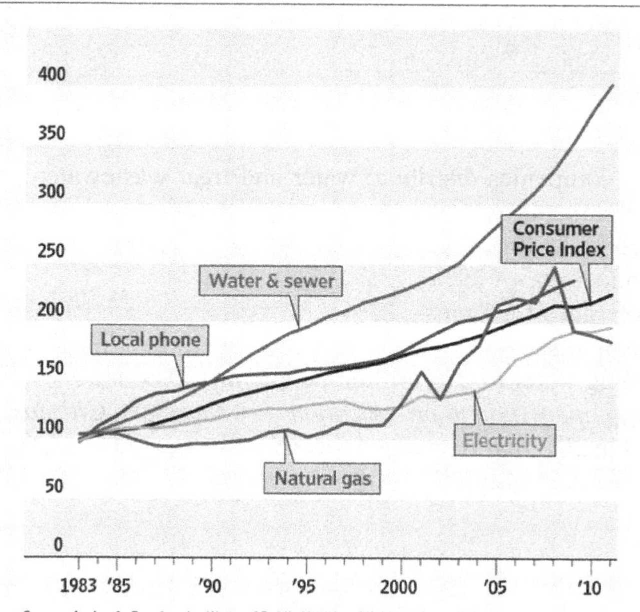

Source: Janice A. Beecher, Institute of Public Utilities, Michigan State University
The Wall Street Journal

France — World Leader in Privatization

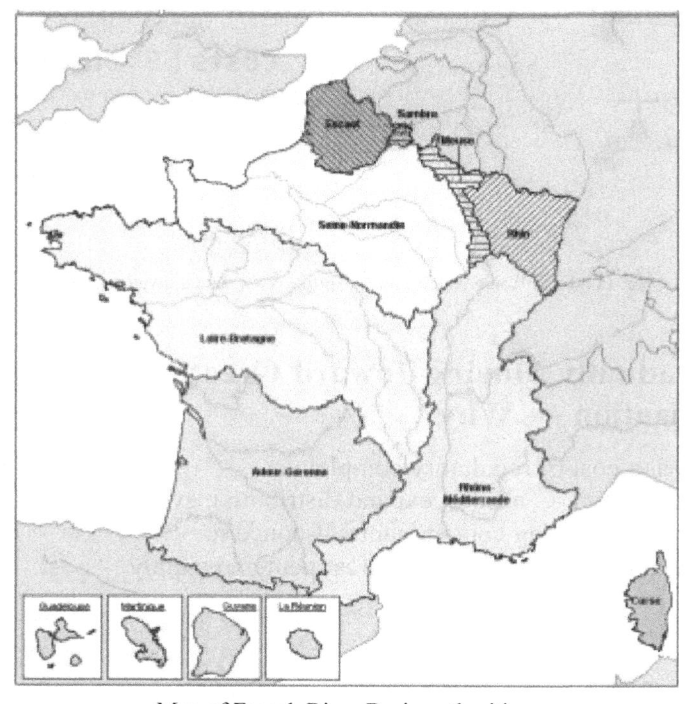

Map of French River Basin authorities.

- *Private* companies distribute water and treat wastewater — 3 enterprises dominate:
  - Veolia — 56%
  - Ondo Suez — 29%
  - Bouygues-La Saur — 23%
  - Companies also operate overseas, in developing countries, providing water to 250 million people in 130 countries.
- *Growing opposition to privatization* — a *European Citizens' Initiative* formed in 2012 has demanded that water be treated as a human right – not as a commodity!

A Case Study of Private Water Service — Bolivia

- **1999: *World Bank* demands country reduce its deficits Bolivia's Congress passes *Drinking Water and Sanitation Law* ending government subsidies, allowing privatization.**
- ***International Water Corp. – a consortium of French and U.S. companies – take control* of water services in Cochabamba; elsewhere: monthly water bills reach $20 (average wage is less than $100/month).**
- **Increases forced poor families to choose between food & water ($20 = cost of feeding family of five for two weeks).**
- **2005: following protests, Bolivia's constitution amended to guarantee *human right to water*, ban privatization.**
- Government lacks money to upgrade water supplies, protect quality – thus, problem of safe, potable access persists.
- **Illustrates consequences of excluding local citizens from water decisions – by private companies AND by government.**

Cochabamba, Bolivia

Protests Elsewhere Over Privatizing Water

Philippines

Nigeria

Argentina

El Salvador

Why Opposition to Privatization?

- Fear of foreign control of water supply — including "*exporting*" *water* via soft drink industry.
- Cost to consumers and failure to deliver on service.
- *Excluding consumers* from operational decisions.
- Belief that access to water is a *human right* — not a marketable commodity.

Solutions? (Pacific Institute, 2012 and Others)

- Require private companies to meet explicit performance measures.
- Require public oversight of operation.
- Permit local governments to regulate company operations.

Bottled Water — Another Privatization Issue

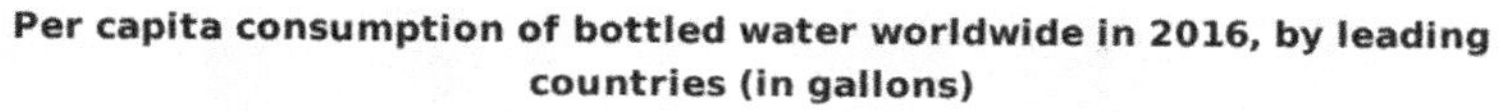

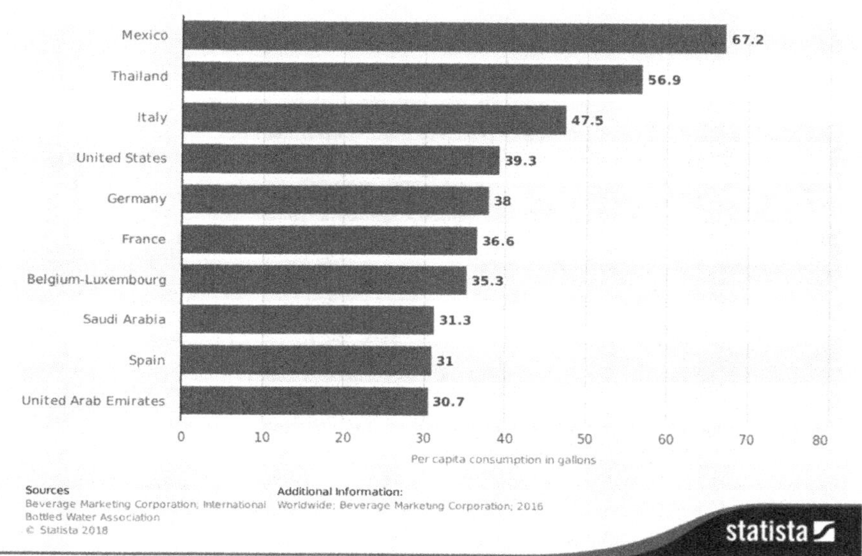

Is Bottled Water *Fair and Equitable*?

- More costly than tap water:
  - Average cost of 60 gal of tap water in US = 13.5 cents.
  - Average cost of 60 gal of bottled water = $48.
- Bottling affects local communities, e.g.,
  - Large corporations (e.g., Evian, Perrier-Nestle, Coke, PepsiCo have purchased water rights in many developing countries.
  - Coke and PepsiCo bottle water under their own labels — citizen-group opposition has arisen due to local water shortages in developing countries.
  - Soft-drink manufacturers have exploited rural sites with local impacts on pristine headwater streams, springs, aquifers.

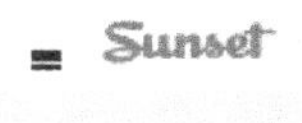

Do You Know Where Your Bottled Water ...

SUNSET + FOOD & DRINK

Do You Know Where Your Bottled Water Comes From? Hint: Partially From Drought-Ridden California

Daniel Miller

The bottled water industry is getting rich off California water. Here's how they're doing it (and how much they're making)

DAKOTA KIM – January 29, 2020 | Updated February 18, 2020

Bottled water companies are feeling the heat in California lately.

From improperly disposing of toxic wastewater to facing accusations of plundering already drought-ridden California's mountain springs and municipal water stores, **the companies that own Arrowhead, Crystal Geyser, Pure Life, Aquafina, and Dasani are under pressure from California residents and environmentalists** (and that's not even including the growing anti-plastic bottle movement **and the study that shows bottled water** contains microplastics).

CG Roxane, the parent company of Crystal Geyser based in Olancha, California, has pleaded guilty to illegally storing and transporting hazardous waste (in this case, arsenic-laden wastewater). The environmental repercussions of their "arsenic pond," as well as of dumping 23,000 gallons of their toxic, potentially carcinogenic water into California sewers, is still unknown. The cost to the company is a $5 million fine, which is not considered a threat to their business.

More Videos From Sunset

Read More

RECIPES

These 12 Pork Chop Recipes Are Anything But (Ahem) Boari...

FOOD & DRINK

Make Cold Brew Like the Pros

FOOD & DRINK

How to Brew Coffee with a French Press

RECIPES

How to Brew Turkish Coffee

Is Bottled Water *Targeted to Minorities*?

- *In California*, 55% of Latinos; 30% of whites drink bottled water as regular source of drinking water.
- One study indicates that black and Hispanic children are 3x more likely to drink bottled water than whites — parents believe it is safer.
- In US, non-whites spend a greater proportion of their income (about 1%) on *bottled water*, compared to 0.4% by whites.
- Many recent immigrants to US *distrust* tap-water — and *marketing strategies by vendors often target poor and minorities*.

Sources: International Bottled Water Association; Public Policy Institute of California Poll; California Legislature — legislative analyst organization; Archives of Pediatric and Adolescent Medicine (2011); Corporate Accountability International.

Is Bottled Water *Healthy*?

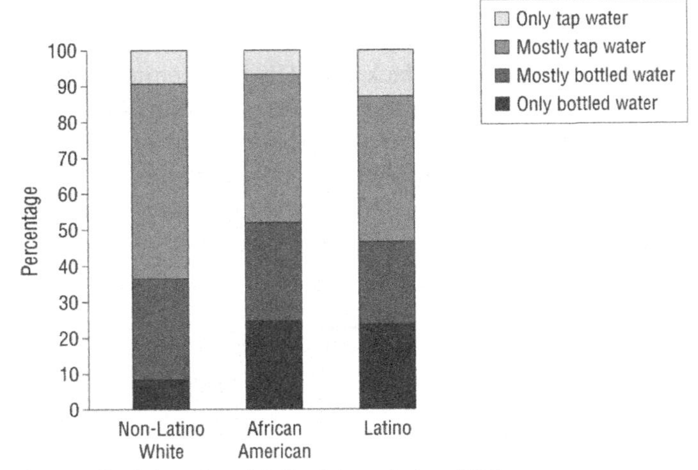

Source: J. of American Medical Association, 2011.

- *Underserved Latinos and African Americans* have been found to have higher rates of bottled water use than non-Latino whites.
- Under-represented groups are more likely to exclusively give bottled water to their children.
- Disparities in use are driven by differences in *beliefs and perceptions* about water.
- Studies have documented elevated bacterial counts in bottled water, and an association between bottled water use and risk of acute diarrheal illness in children.
- Use of bottled water in place of tap water may lead to *inadequate fluoride intake* for children — implications for oral health.

Is Bottled Water *Safe*?

Key Differences Between EPA Tap Water and FDA Bottled Water Rules

Note: To print this chart, in the Print dialogue box choose Properties and Paper and set to Legal and Landscape and click OK; under Print Range choose "from 1 to 1" and click OK (this will print one page and lock in settings); then use Print Preview to determine which page(s) to print.

Water Type	Disinfection Required?	Confirmed *E. Coli* & Fecal Coliform Banned?	Testing Frequency for Bacteria?	Must Filter to Remove Pathogens, or Have Strictly Protected Source?	Must Test for *Cryptosporidium, Giardia,* Viruses?	Testing Frequency for Most Synthetic Organic Chemicals?	Operator Must be Trained & Certified?	Must Test for and Meet Standards for Asbestos & Phthalate?	Must Use Certified Labs to Do Testing?	Must Report Violations to State, Feds?	Consumer Right to Know About Contamination?
Bottled Water	No	No	1/week	No[a]	No	1/year	No	No	No	No	No
Carbonated or Seltzer Water	No	No	None	No	No	None	No	No	No	No	No
Big City[b] Tap Water (using surface water)	Yes	Yes	Hundreds/ month	Yes	Yes	1/quarter (limited waivers available if clean source)	Yes[c]	Yes (though limited waivers available if clean source)	Yes	Yes	Yes
Small Town[d] Tap Water (using a well)	No (though new rule in 2002 will require if needed)	Yes	20/month	No (unless subject to surface contamination)	No	1/quarter (waivers available if clean source)	Yes[c]	Yes (though waivers available if clean source)	Yes	Yes	Yes

a. FDA requires state or local approval of bottled water sources, but there is no federal definition or control of what may be a bottled water source; the FDA "approved source" requirement thus has been called a "regulatory mirage."

b. Big city refers to city system serving 100,000 people or more. A big city using only wells would have to comply with all requirements noted for a surface water-supplied city, except that if its wells were not under the influence of surface water, it currently would not have to disinfect, filter, or test for *Cryptosporidium, Giardia,* or viruses. A new rule for such groundwater-supplied systems must be issued in 2002, which may require some cities using wells to disinfect or filter and do additional microbial monitoring.

c. The Safe Drinking Water Act Amendments of 1996 require states, subject to EPA guidelines, to train and certify operators of all public water systems. EPA's rules to implement this provision are required to be issued by February 1999.

d. Small town refers to a town of 20,000 people. Such a small town using surface water would have to comply with all the same requirements noted for a large city using surface water, except the monitoring frequency for coliform would be 20/month, and there currently are no *Cryptosporidium, Giardia,* or virus monitoring requirements for small towns.

Source: NRDC

Alternatives to Conventional Bottled Water

Water-On-the-Go:

- New York City Department of Environmental Protection offers portable fountains for drinking or *filling water bottles* and to "make NYC water easily accessible to attendees at your event". (Like UCI's hydration stations.)

New York City has some of the best tap water in the world. Grab your reusable bottle and drink up!

- Each day, more than 1 billion gallons of fresh, clean water are delivered to NYC from pristine reservoirs in the Catskill Mountains.

- New York City drinking water is world-renowned for its quality. The Department of Environmental Protection performs more than 900 tests daily, 27,000 monthly, and 330,000 on an annual basis from up to 1,200 sampling locations throughout New York City. This work is in addition to 230,000 tests performed in the watershed.

- Save money! NYC water is a great deal. At approximately one penny per gallon, it is about 1,000 times less expensive than bottled water!

Section 5

Water Quality — Impacts to Health, Environment, and Well-Being

Elections promise to steer direction of water policy on Capitol Hill

The results of this week's presidential and congressional elections will carry profound consequences for the direction of the nation's drinking water policy over the next two years.

Currently Democrats hold a 17-seat majority in the House of Representatives and Republicans hold a narrower, three-seat edge in the Senate. Most forecasters expect Democrats to easily maintain control of the House while having a strong chance taking over the Senate as well.

Should Democrats come out victorious on Capitol Hill and in the White House, they are expected to move aggressively to enact a far-reaching agenda that will carry numerous implications for drinking water policy. Legislation to address global climate change would be high on the to-do list, as well as efforts to revise the contaminant regulatory process under the Safe Drinking Water Act and to mandate drinking water regulations for contaminants in the PFAS family.

A Democratic sweep would make consideration of a major infrastructure bill almost certain next year, likely with a significant drinking water and wastewater component. A Democratic-majority Congress could also move quickly next year to enact a major package of economic stimulus legislation to counter the COVID-19 pandemic, especially if Joe Biden is in the White House. Such a package would probably include funding to cover the water bills of some low-income customers but could also prohibit water service disconnections related to nonpayment for the duration of the pandemic, another Democratic priority.

Additional insights into the water policy implications of the elections can be found in this month's edition of AMWA's Congressional Report.

Following Election Day, Congress will return to Washington to carry out a "lame duck" session to conclude the 116th Congress. Among the top orders of business will be enacting FY21 spending legislation before December 11 to avoid a federal government shutdown. Lawmakers will also aim to enact a Water Resources Development Act reauthorization before year's end.

Water Quality — Threats and Challenges

- Water contamination threatens people's health and the environment worldwide. *Nearly 800 million people lack sanitation* and over 80% of wastewater is un-treated.
- While *point source* discharges are better controlled than in the past, *non-point runoff* is poorly regulated (nutrients and chemicals from paved surfaces, farms):
    - *New threats are growing*: personal care products and pharmaceuticals — difficult to detect, treat.

Complex pathways of water pollution.

What's Point Source Pollution?

- The US and other industrialized countries have lowered point source discharges >90% since 1970 by imposing discharge regulations and permits.

What's Non-Point Source Pollution?

- In US, since 2009, Clean Water Act regulates municipal storm sewer systems and their drainage basins, as well as construction, any other activities causing *runoff.*
- Mitigation remains difficult — most cities are out of compliance. In 2014, US Supreme Court held that Los Angeles County is responsible for untreated storm water — in 2018, *it may have found a solution.*

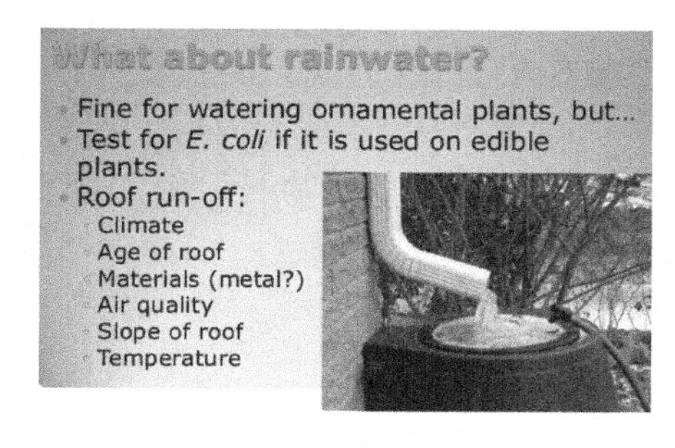

Recent Reforms — Los Angeles County

Measure W, a parcel tax of 2.5 cents a square foot of "impermeable space", earned 70% of vote, November 2018.

The tax . . . will help cities across Los Angeles County meet their obligations under Clean Water Act. Supporters said it would also help make the region *more water* resilient in the face of drought and climate change.

Revenue is supposed to pay for projects to improve water quality and also increase water supply for parks and wetlands (i.e., non-potable uses).

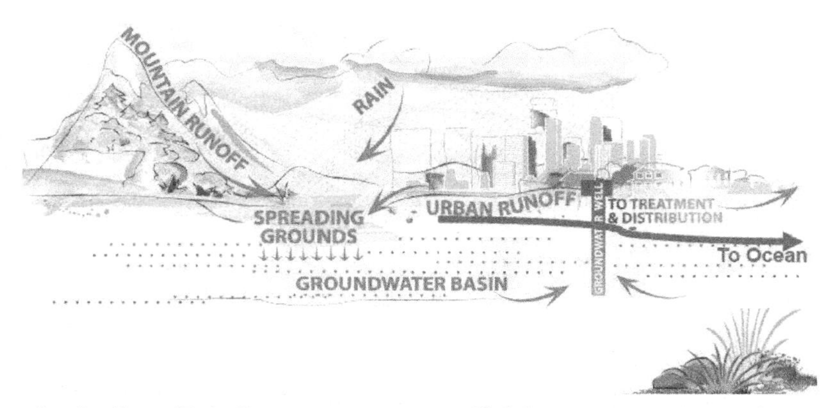

Los Angeles Times "L.A. County stormwater tax officially Passes", By NINA AGRAWAL November 30, 2018.

What Australia can Teach us — Managing Rainwater Runoff

- During millennium drought (1990s–2000s):
  - Local governments adopted *development consents* to encourage installation of rainwater tanks connected to roofs to provide water for gardens, toilets, clothes and car washing; *a method to intercept stormflows.*
  - As further incentive for adopting above- and under-ground storm-water storage/on-site detention, developers and home owners given *rebates* on water and sewer charges.

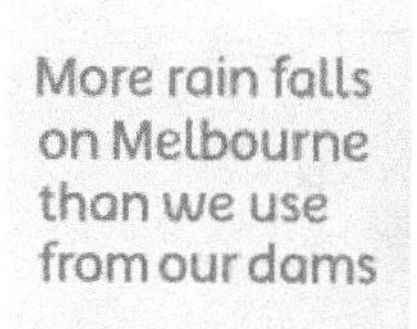

Constructed wetland.

Melbourne beach closure — January 1, 2017.

- Wetlands and bio-filters treat storm-water for non-potable uses, or for pre-treatment for later potable use.
- Lessons: stormwater harvesting encompasses a number of land use and urban planning and design issues.

How Water Pollution Became a *Recognized* Problem

- In 1832, New York City suffered an epidemic of cholera that killed 3,515 out of a population of 250,000. (Equivalent death toll in today's city of eight million would exceed 100,000.)
- Causes? Deplorable sanitation and contaminated drinking water. Led directly to upgrades in local water supply (i.e., reliance on deep wells, imports of water from Catskills) to avert communicable diseases.
- In later 19th century, it led to the birth of what became the field of "sanitary engineering" and the construction of wastewater treatment systems.

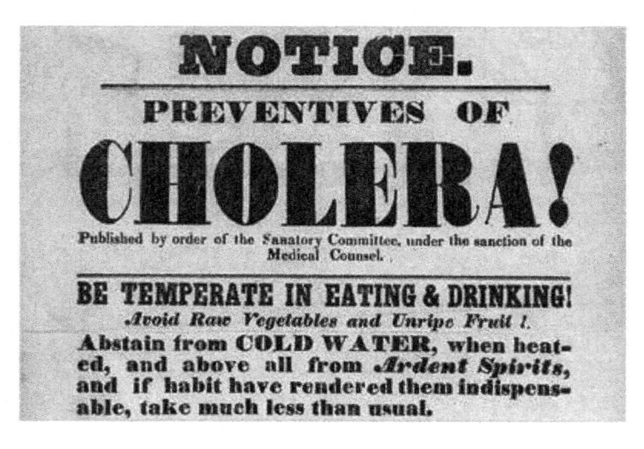

In some Places — Cholera from Pollution is Still a Problem

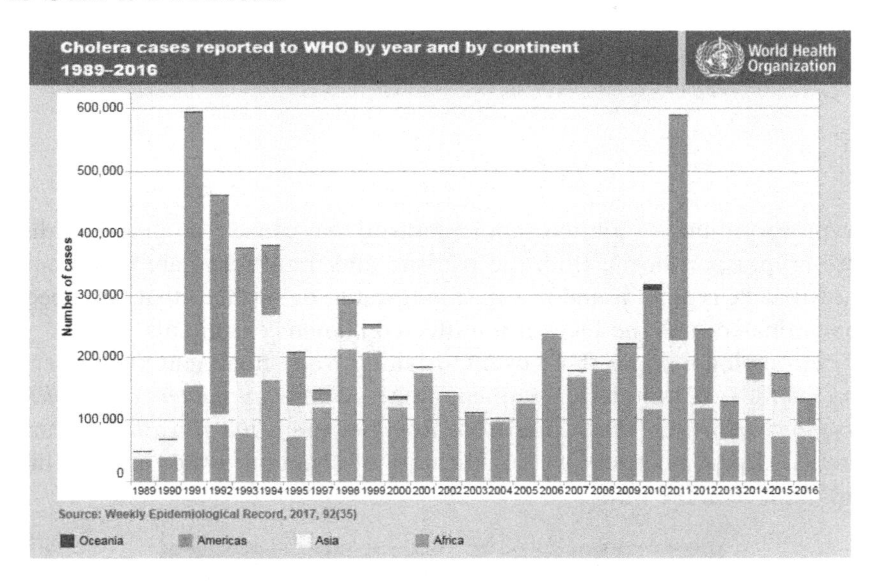

A Legacy of Cholera Epidemics — Sanitary Engineering

- Wastewater treatment systems generally use a combination of methods to remove contaminants: e.g.,
    - Mechanical
    - Biological
    - Chemical
- Releases are unsafe for *human* consumption – only safe enough to be released into waterways under *Clean Water Act* (1972).
- Further treatment needed to remove heavy metals, some pathogens, *perfluoroalkyl* substances – under conditions of *Safe Drinking Water Act* (1974).

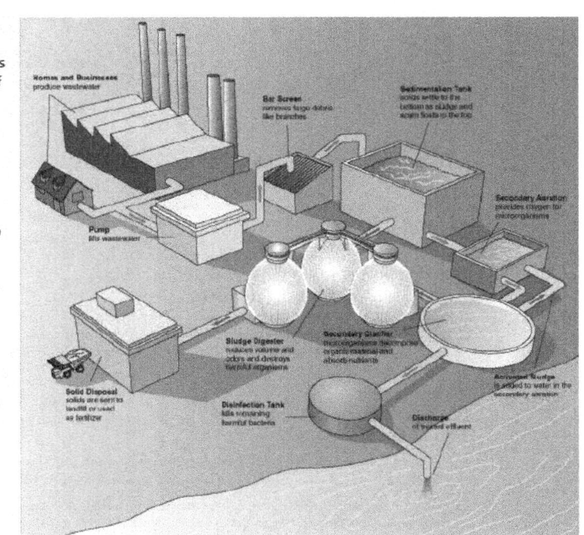

Pollution and Public Safety — More Recent History

Milwaukee Remembers Cryptosporidium Outbreak of 1993

After deadly drinking water contamination a quarter of a century ago, Milwaukee now leads the field in water testing

By **Jared Raney**

May 22, 2018

Cryptosporidium — a microscopic parasite that causes the diarrheal disease cryptosporidiosis. Both the parasite and the disease are commonly known as "Crypto". found in soil, food, water, or surfaces that have been contaminated with the feces from infected humans or animals.

Following a major flood event, drinking water treatment plant operators lost full control of the treatment process *that would detect and isolate Crypto, which was thus allowed to break through the filtration system.* They also paid no attention to indicators of changing water quality they had in place at the time.

It's every treatment plant's worst nightmare: the people you serve getting sick from your plant's water. The most potent example of this is the Milwaukee *Cryptosporidium* outbreak of 1993. Over 400,000 people were affected and more than 100 died.

Coming back from such a disaster takes fortitude and commitment — and for the city of Milwaukee, over $90 million in immediate treatment upgrades. To date, the city has spent $508 million in water infrastructure upgrades, not all related specifically to *Cryptosporidium*, but all in an effort to ensure their citizens are protected.

The outbreak

This April marked 25 years since the infamous epidemic, which was likely the largest waterborne illness outbreak in U.S. history.

It took several days to identify the cause and source of the illness running rampant through Milwaukee, partially due to the lack of adequate testing facilities and an unfamiliarity with the contaminant.

A boil notice was issued for seven days, and they shut down the treatment plant that the *Cryptosporidium* had been linked to. The notice was only lifted after the plant had been thoroughly purged.

For Discussion

- Have we made good progress in addressing water pollution? Why, why not?
- Are the underlying causes of water pollution being adequately addressed? Yes or no.

Unconventional Threats — Contaminants of Emerging Concern

- **Contaminants-of-concern:**
  - **Remnants of prescription and over-the-counter drugs, cosmetics, personal care products, neutraceuticals (vitamins), pain-relievers, antibiotics, veterinary drugs, flame retardants, micro-beads**
- Major health/environmental concerns:
  - **Enter the environment after we (or domesticated animals) use them.**
  - **Difficult to detect or remove – many water systems not equipped to treat them.**
  - **Risks widespread – disrupt aquatic and human endocrine systems; increase bacterial resistance to antibiotics; are mutagenic, carcinogenic.**
  - **Number of chemicals growing; not all have been tested for toxicity.**
  - **Bio-accumulative & persistent – e.g.,** Per- and Polyfluoroalkyl substances or **PFAS**

WHAT CAN WE DO?

- **Dispose of personal care products/drugs in trash, or use community "take back" programs to dispose.**
- **Only flush *if* labels say so – most COCs enter water supply through disposal in toilets, sinks.**

Contaminants of Concern — In Our Backyard

Methods

The exposure and potential effects of emerging contaminants were investigated on fish living areas near the four largest ocean POTW discharges in southern California. The study areas included the Los Angeles County Sanitation District (LACSD), City of LA, Orange County Sanitation District (OCSD), and City of San Diego outfalls, which represented range of effluent characteristics and levels of legacy contamination in sediments. A reference area at a similar depth offshore of Dana Point was also sampled.

The study area included four sites near the largest offshore discharges of municipal wastewater and a reference site near Dana Point.

- 2016 *Southern California Coastal Water Resource Project* found many contaminants in wastewater plant outfalls, generally at concentrations of 1–10 micrograms per liter.
- Aquatic life is exposed to a wide variety of emerging contaminants, even after 100- to 1000-fold dilution of wastewater effluent in the ocean.
- Future monitoring of fish tissue and sediments in rivers flowing into coastal waters is warranted.

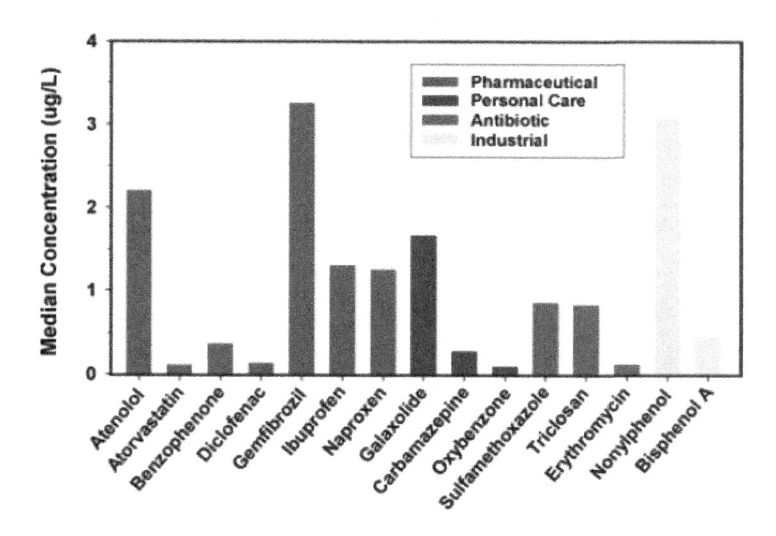

- *Per- and Polyfluoroalkyl substances (PFAS) a group of >3,000 chemicals that repel oil, water.*
- Used in non-stick coatings, carpets, rain-gear, paints, polishes, waxes, cleaning products, food packaging. Firefighters and military use them in fire-suppressing foam.
- Persistent in environment and human body — accumulate and have a 2–4 year half-life in humans.
- US Environmental Protection Agency (USEPA, 2020) — Perfluorooctanoic acid (PFOA) and Perfluorooctanesulfonic acid (PFOS) can cause reproductive and developmental, liver and kidney, and immunological effects in lab animals and tumors; increased cholesterol levels among exposed populations; other effects include:
  - low infant birth weights,
  - effects on the immune system,
  - cancer (PFOA),
  - thyroid hormone disruption (PFOS).

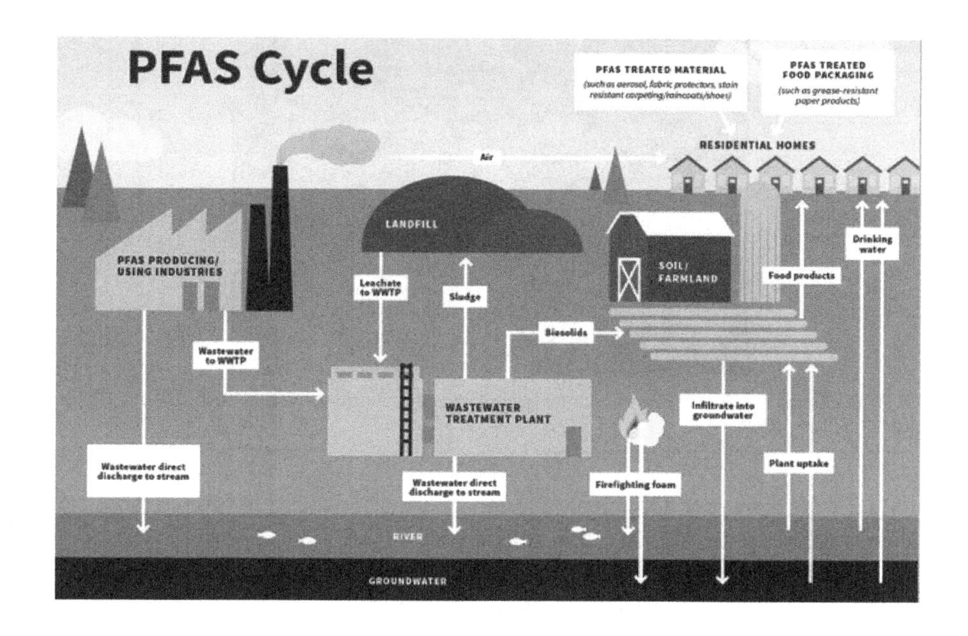

Orange County Water District (OCWD), which serves 2.5 million county residents, will see nearly a third of its 200 groundwater wells shut down by end of 2020 because of the presence of toxic PFAS.

OCWD manages a groundwater basin that provides 77% of the water used by 19 member agencies, which pump the water from wells in central and north Orange County.

https://www.ocwd.com/what-we-do/water-quality/pfas/.

OCWD researcher loads ion exchange resins into a treatment system to filter out PFAS toxins.

WATER UCI project: Objectives of this proposed project are to (1) develop a methodology to identify the sources of PFAS in the sewer-shed of a Wastewater Treatment Plant (WWTP) in Orange County, that can be replicated elsewhere for PFAS or other contaminants; and (2) work with a WWTP ... to implement the methodology, with the goal of identifying the primary sources of PFAS to the Plant's sewerage system. Additional studies will be conducted to identify specific products leading to large PFAS release from residences.

Flint, Michigan — "It's Ridiculous we have to Live in Such a Way"

- Quote from Colette Brown, Flint, MI native who stopped drinking tap water.

The Saga of Flint — Lead in Our Drinking Water Flint

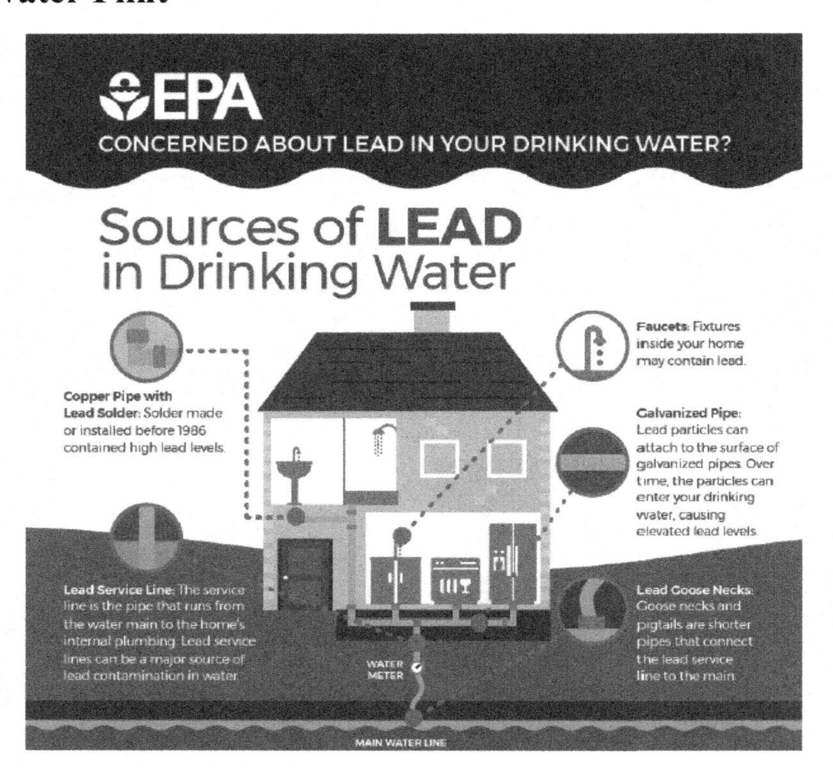

Timetable to a crisis

- *April 2014*: Flint cancels agreement with Detroit's municipal water supply system — joins regional authority to build pipeline to Lake Huron — done to save $18 million.
- As interim measure, city's emergency manager, appointed by Michigan's governor to restore fiscal solvency, switches supply to Flint River.
- Almost immediately, residents complain of "discolored, foul-tasting, awful smelling water".
- Fall 2015: discovery that corrosive water (19 times more corrosive than Lake Huron) leached lead from century old galvanized iron-pipe system.

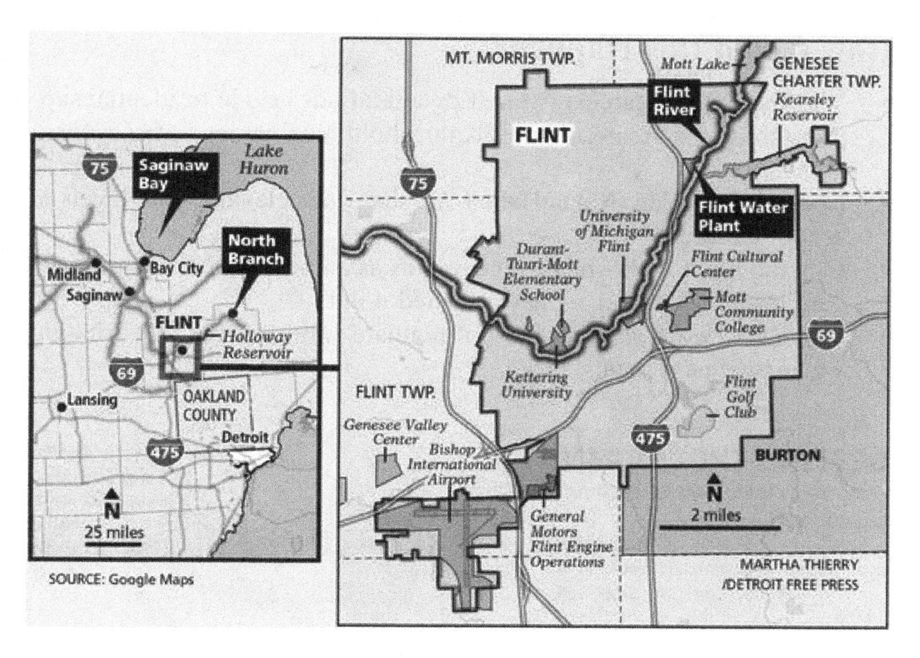
Lake Huron
Saginaw Bay
North Branch
Midland
Bay City
Saginaw
FLINT
Holloway Reservoir
Lansing
OAKLAND COUNTY
Detroit
25 miles
SOURCE: Google Maps
MT. MORRIS TWP.
Mott Lake
GENESEE CHARTER TWP.
Kearsley Reservoir
Flint River
FLINT
Flint Water Plant
University of Michigan Flint
Durant-Tuuri-Mott Elementary School
Flint Cultural Center
Mott Community College
Kettering University
FLINT TWP.
Genesee Valley Center
Bishop International Airport
Flint Golf Club
BURTON
General Motors Flint Engine Operations
2 miles
MARTHA THIERRY /DETROIT FREE PRESS

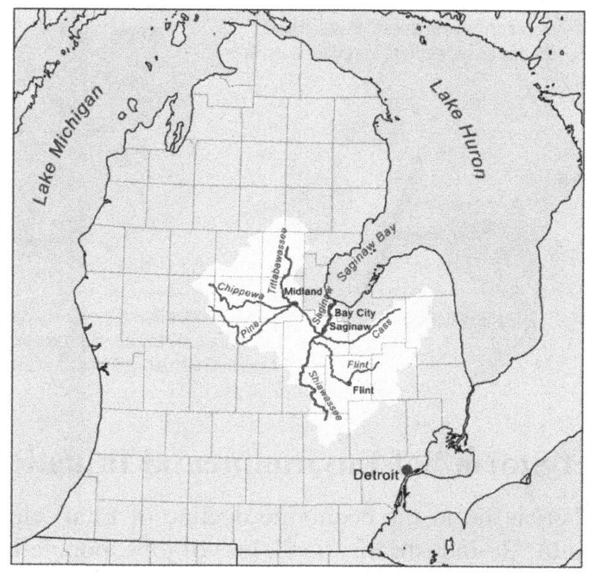
Lake Michigan
Lake Huron
Saginaw Bay
Tittabawassee
Chippewa
Midland
Pine
Saginaw
Bay City
Saginaw
Cass
Flint
Flint
Shiawassee
Detroit

How Could this Happen?

- Virginia Tech researcher (Marc Edwards) finds lead in residential supplies six-times greater than EPA threshold level under *Safe Drinking Water Act*.
- Local pediatrician (Mona Hanna-Attisha) finds elevated lead levels in school children.
- City did not use phosphate compounds as anti-corrosion agent to prevent leaching of lead — state claimed it did.
- *Safe Drinking Water Act* did not regulate water after it left drinking water treatment plants!

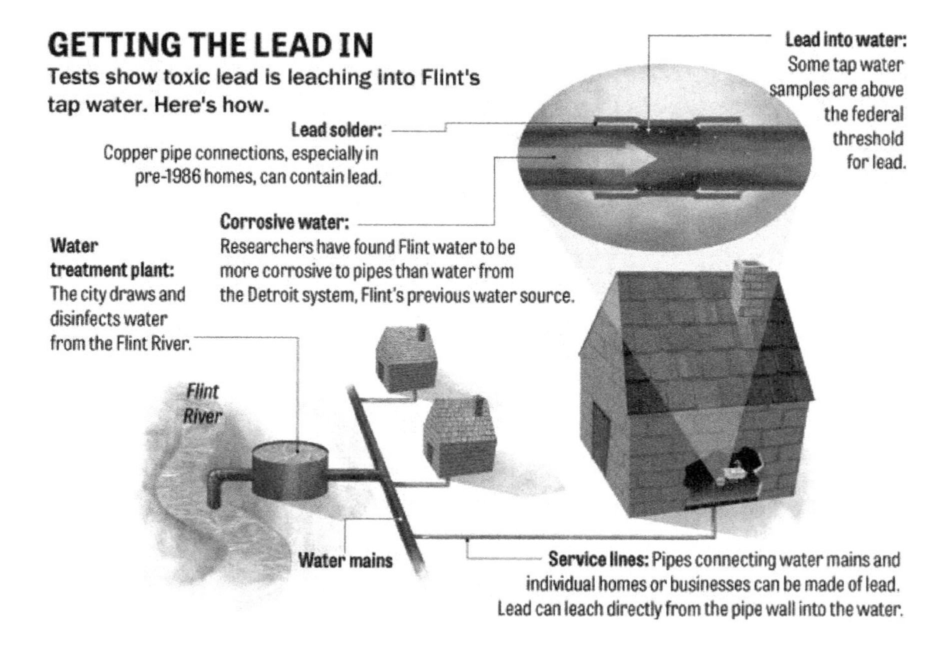

A "Perfect Storm" of Environmental Injustice

- Roots of crisis lie in the economic decline of Flint: closure of GM plants led to "de-industrialization". loss of jobs, population, tax base, utility rate base.
- While city's population declined from 200,000 (mid 1960s) to fewer than 100,000 by 2015, cost of maintaining municipal water supplies did not decline.

  - Operation and maintenance of distribution and treatment systems don't decrease from "under-use".
- Efforts to save money failed to consider corrosion risk, AND was made without real political accountability — city's emergency manager was not elected.
  - Citizen complaints ignored.
  - 40% of residents live below poverty line — chronically disempowered from decision-making.
- *Thousands of children ingested high levels of lead* — mental and physical health problems may be latent for years; families will bear high costs-even with proposed federal aid and legal remedies.

Nationwide — Drinking Water Quality Problems Widespread

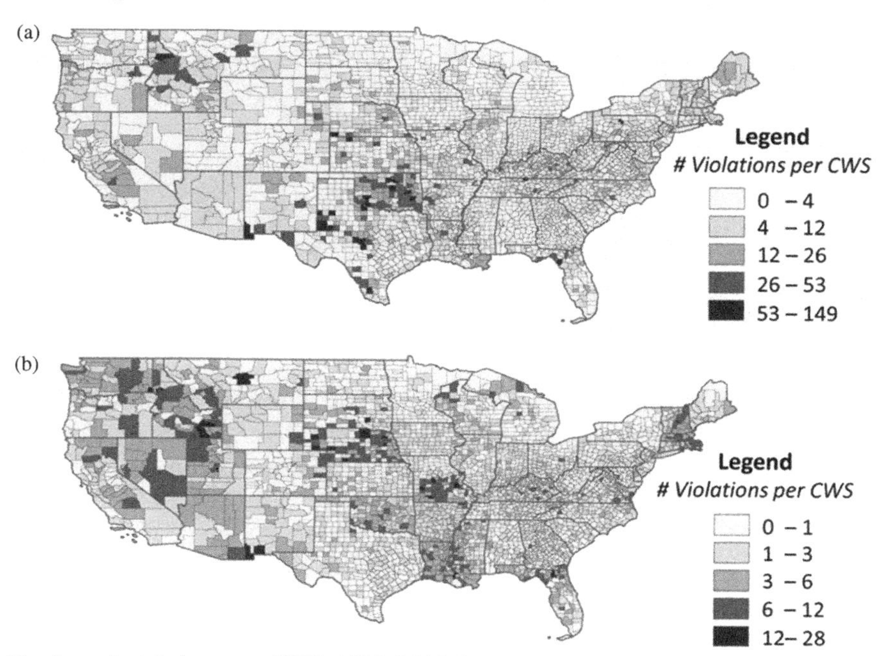

Number of violations per CWS, 1982–2015, by county. (A) Total violations. (B) Total coliform violations "National trends in drinking water quality violations", Maura Allaire, Haowei Wu, and Upmanu Lall. PNAS February 27, 2018 115 (9) 2078–2083; https://pubmed.ncbi.nlm.nih.gov/29440421/.

Nutrient Pollution — Need for Innovation

- Runoff has degraded Great Lakes, Gulf coast, Atlantic seaboard, elsewhere.
- Levels of toxins from algae has forced bans on tap water use, even for bathing, for those with "compromised immune systems".
- Excess runoff from fertilizers, livestock feeding, and sewage treatment plants is cause.

Gulf of Mexico dead zone is an area of hypoxic (less than 2 ppm dissolved oxygen) waters at the mouth of the Mississippi River. Varies in size, but can cover up to 7,000 square miles. Most of the nutrients come from farming states in the Mississippi River Valley.

Nutrient Trading — Is it a Solution?

Water table control structure to reduce excess nutrients.

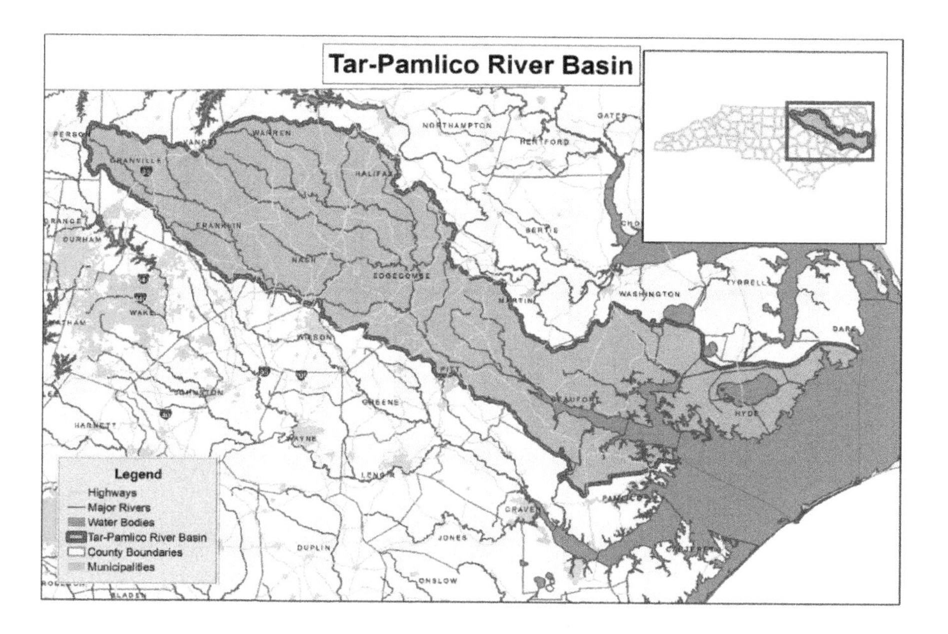

- North Carolina's *Tar-Pamlico River* basin — nitrogen and phosphorus runoff from farms and point discharges from water treatment plants in cities led to low dissolved O_2, fish kills, vegetation loss.
- 2000: state adopted current "nutrient trading framework." Cities pay farmers to install water control structures to treat cropland, avert run-off, plant nutrient feed crops. Goal: reduce nitrogen loadings by 30%; and no increase in phosphorous.
  - *Places mandatory* controls on agricultural non-point sources, urban stormwater, nutrient management, riparian buffers.
  - By 2016: farmers achieved significant reductions in nitrogen and phosphorous — met targeted goals.
  - HOWEVER — estuary is still classified as "impaired" and not meeting *nitrogen loading reduction* goals.

Nitrogen Reductions by Agriculture

Figure 1. Collective Cropland Nitrogen Loss Reduction Percent 2001 to 2016, Tar Pamlico River Basin.

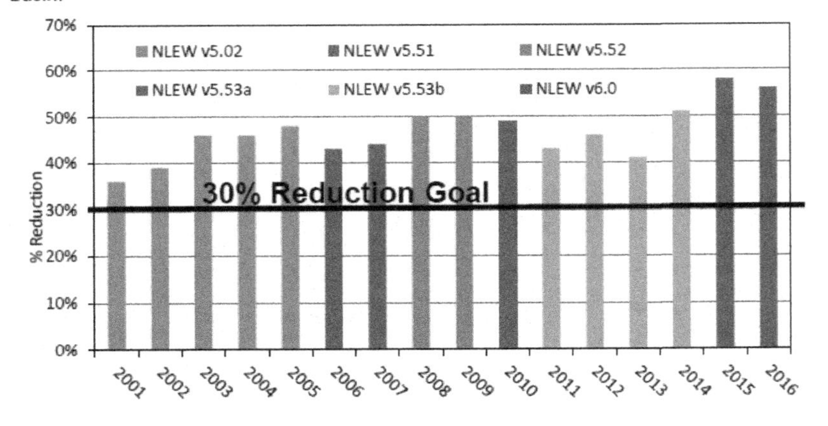

Nitrogen Loadings (Concentrations) — Pamlico Estuary

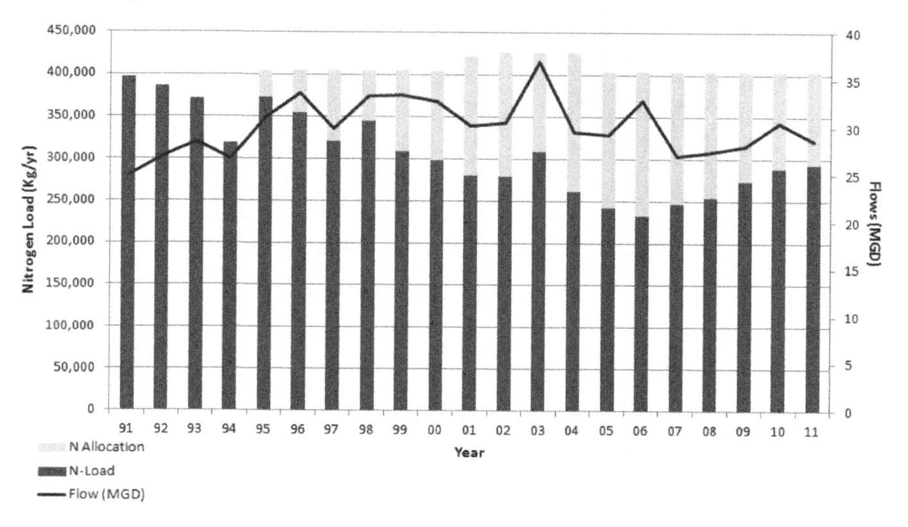

- Nutrients remain a persistent problem — possible reasons are changes in land use, more CAFOs, but NOT urban WWTPs (NC Sea Grant program 2017).
- Partly influenced by "inputs" — partly a problem of "dilution" (i.e., flow); also partly caused by airborne pollutants (nitrogen oxides).

Water Pollution is a Global Problem — Rhine River

Chemical complex — Wesseling, Germany.

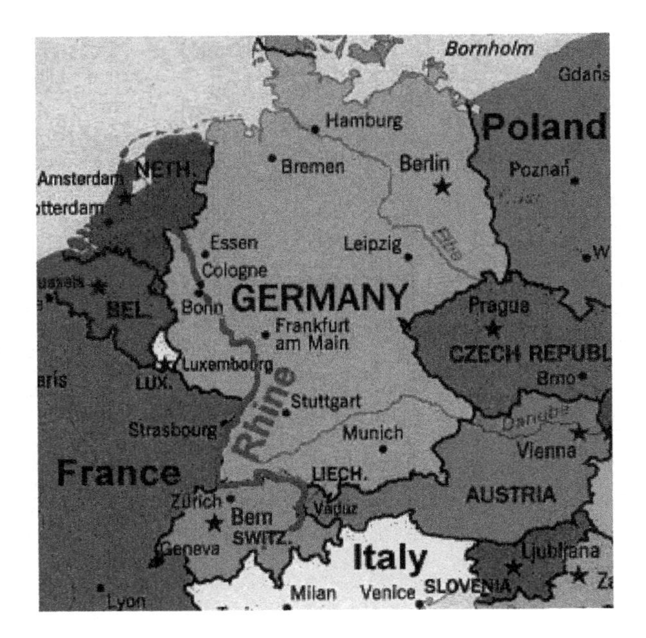

Industrial pollution near Cologne.

Dead eels following Swiss chemical spill — 1986 near Basel.

What Rhine Basin Countries are Doing

- 1999 — basin countries adopted International *Convention on Protection of the Rhine*, committing them to common water quality goals, restoration of North Sea fisheries.
- *Have reforms helped*?
  - Improvements attributed to *public outrage*, demands to *improve river as an amenity*.
  - Point source pollution gradually diminishing — heavy metals lessening, oxygen levels increasing; salmon, eel, pike, trout re-appearing.
  - Agricultural runoff still prevalent; contaminants of concern a growing concern (as in USA).
  - Member states are reluctant to grant ICPR *independent enforcement power*.

Conclusions

- Water pollution usually managed through *end-of-pipe remedies*: complex, expensive.
- An alternative? *Pollution prevention*, also known as *precautionary policy*; e.g.,
  - Avert pollution by mandating safer products.
  - Encompass *life cycle costs* of products in regulations and in pricing.
  - Prevent or intercept *environmental pathways that* affect health and welfare — remember, everyone is "downstream" of a hazard.

Figure 12. Deep-sea litter, including plastic beverage bottles, 20 km off the Mediterranean coast at a depth of 992 m. *Image courtesy of Francois Galgani/Ifremer.*

20 million tons/year of plastic waste enter our oceans.

Section 6

Sources of Water Conflicts — Diversion, Depletion, and Degradation

Sources of Water Conflicts — Diversion, Depletion, and Degradation

- *Diversion* — taking water from one basin and transferring it to another *without return flow. Imposes adverse effects on people and the environment.*
- *Depletion* — using up available ground-or surface water *without replenishment. Imposes adverse effects on people and the environment.*
- *Degradation* — reducing the healthy functioning of a water body through pollution and other changes to its environment. *Adverse social and environmental impacts.*
- These problems may be at a "tipping point" — *impacts* may be irreversible, even if the activities ceased.

Aral Sea — Diversion and Depletion

- World's fourth largest saline lake until 1970s (26,000 mi^2).
- 1960s — Soviet Union diverted Syr Darya and Amu Darya to irrigate cotton, other crops.

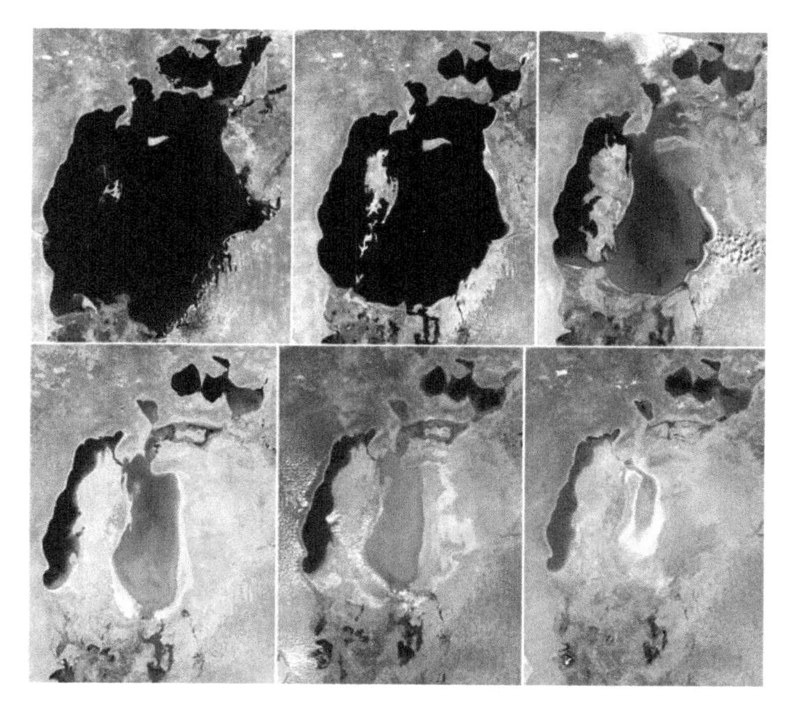

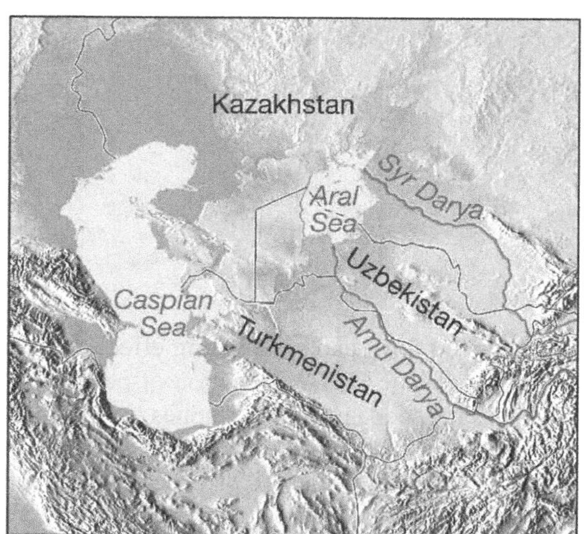

Top left view — mid 1960s; remainder taken at 10 year intervals; last taken 2014.

Aral Sea in Crisis

"...salt and dust laden air have had a damaging effect on the health of the people, and animal and plant life. Introduction of pesticides into the rice and cotton fields, and the seepage of the residues back into the rivers, has heavily contaminated the water for those communities living along the banks of the rivers flowing into the Sea... Diseases like anemia, cancer and tuberculosis, and allergies are frequent. (and) an increase typhoid fever, viral hepatitis, TB, and throat cancer in many areas as high as three times the national average". — Aral Sea website — maintained by NGOs in Kazakhstan.

Aral Sea and Water Resource Policy

- After diversion, most water soaked up by desert.
- Aral normally receives 20% of water from rain, 80% from Amu Darya and Syr Darya rivers — sustainable as long as inflow = evaporation; diversion caused imbalance
- Salinity rose 10-fold in Southern Aral — killing most fish species.
- Government proposed diverting Siberian rivers to replenish Aral — their own scientists said NO.

Arrows represent contemplated diversions from Arctic rivers into Aral Sea.

… An Update

- 2005 — World Bank and Kazakhstan built Kok-Aral dam to raise N. Aral by 13 feet, deep enough to drop salinity, allow native fish to repopulate sea. $85 million project also improved irrigation structures upriver.
- 2006 — UNESCO designates Aral Sea "World Heritage site" to bring greater attention to problems.
- 2017 — N. Aral has been partially restored.

Humans Killed the Aral Sea. Now, It's Come Back to Life, Wired magazine, 2/14/17

"The water is back — it's like a fairytale", says French photographer Didier Bizet, who documents the turnaround in his ongoing series Aral Dreams. "Suddenly, in the Aral Sea, life is coming back". Bizet spent two weeks joining fisherfolk on their motorboats as they went to sea each morning to cast their nets and again each afternoon to haul in carp, pike, and perch. Bizet also met townspeople — many from families who've lived in the area for generations — in local cafes or at their homes. Over the past five years, many people have returned to the area, and everyone feels hopeful for the future.

The Salton Sea — Playground or Battleground?

- California's largest lake. Provides critical habitat for >400 species of migratory birds, including many endangered species (e.g., White Pelicans).

- Farming, water transfers to cities, climate change are: shrinking lake, contaminating return flow, threatening habitat, increasing salinity, degrading air quality.

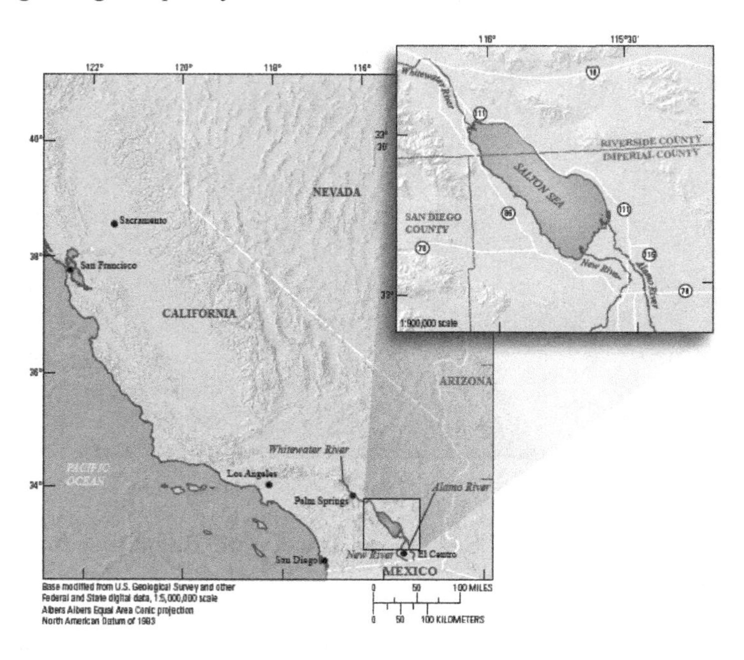

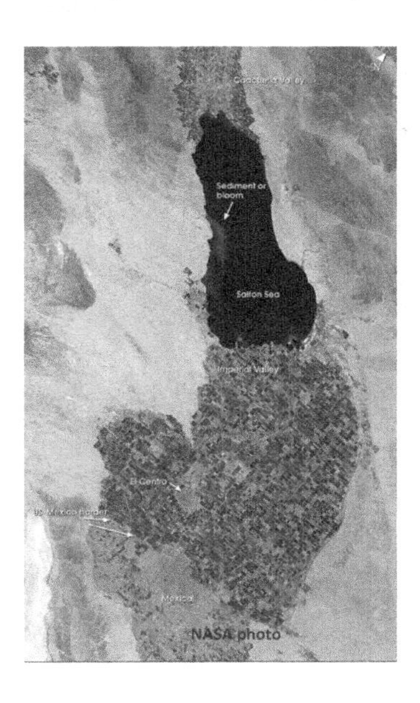

Nature Rules — from "Sink" to "Sea" and Back Again

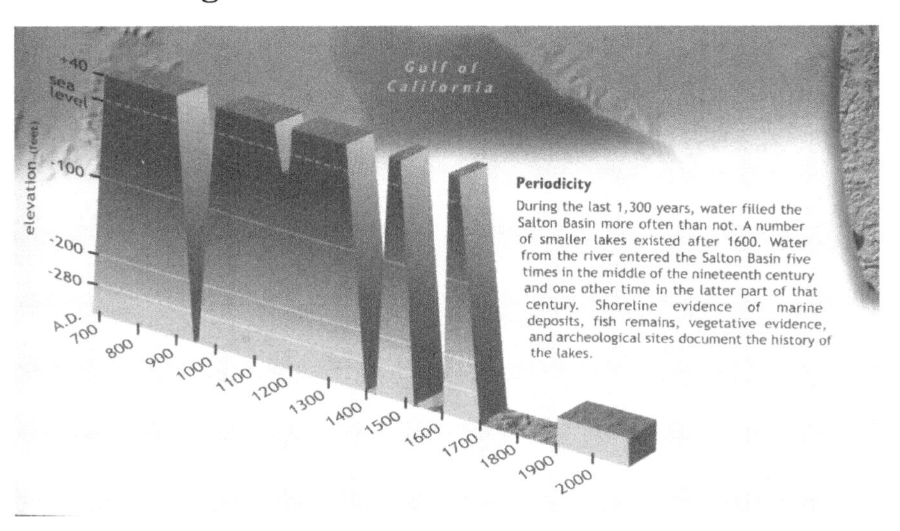

- Salton basin periodically flooded by Colorado River (Lake Cahuilla was one result); forming a naturally fertile valley.
- *Late 19th Century*: investors sought to irrigate basin, sell land parcels, create an *inland empire* of agriculture — thus, *Imperial Valley* born.

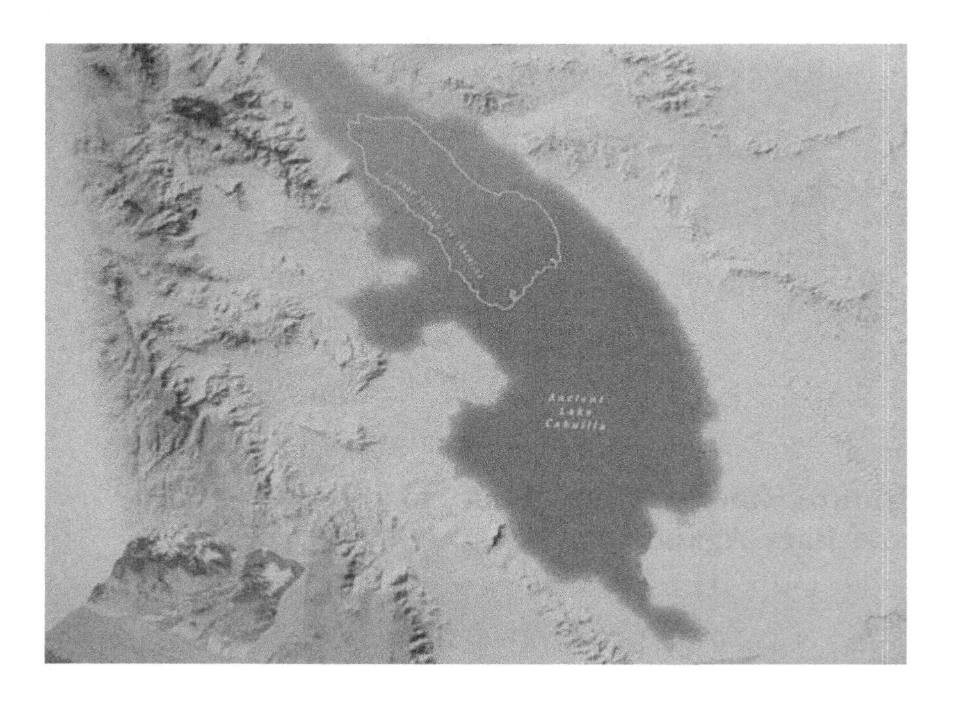

From Human Diversion to Catastrophic Flood

- 1904–1905: Colorado River broke through irrigation works and flooded Imperial Valley for two years.

Can it be Restored?

- TODAY: Dust contains silicates, toxic and caustic salts, and heavy metals such as cadmium, arsenic and selenium.

Edited by **Jennifer Sills**

Salton Sea: Ecosystem in transition

Each year at least a billion birds migrate down the west coast of North America along what is known as the Pacific Flyway (*1*). For the aquatic species, large, open bodies of water are important oases. For more than a century, the Salton Sea in southern California has been a vital component of this flyway for fish-eating birds (*2, 3*), including pelicans, cormorants, skimmers, mergansers, herons, egrets, terns, and gulls. In California, 96% of the historic wetlands have been lost to agriculture and development (*4*). The delta of the Colorado River, in Mexico just south of California, was once a verdant wetland but is now shrunken and dry. In most years, not a drop of water in the river reaches the Gulf of California in Mexico (*5*).

In January 2018, mitigation water that had been flowing to the Salton Sea to maintain its level and salinity began to be diverted, which will cause the sea to shrink in volume and increase in salinity (*6, 7*). These changes will eliminate fish. The Salton Sea will shift from an ecosystem in which fish are the trophic level on which birds feed, to one in which birds feed on invertebrates.

It is far from clear how birds of the American West will respond to this shift (*4*). It may lead to fewer bird species overall, but some species will likely increase in number, such as the eared grebe, ruddy duck, northern shoveler, California gull, and ring-billed gull. Populations of saline lake specialists, such as black-necked stilts, Wilson's and red-necked phalaropes, and American avocets, will likely increase as well.

The human-induced ecological transition of the Salton Sea has profound implications for continent-wide avian populations (*8*), as well as for human communities exposed to toxic dust rising from the drying shores (*9*). Currently, the State of California has not adequately addressed the ecological consequences and public health impacts of a shrinking and salinizing sea (*10*). It is important to plan for an ecosystem-wide transition that minimizes the impacts on birds and on the hundreds of thousands of human inhabitants living adjacent to the shrinking sea. Steps toward restoration include building and maintaining suitable habitats on the periphery of a smaller Salton Sea that will reduce dust emission and create more habitats for birds; monitoring the quality of shallow water habitats; and minimizing the release of dust from newly exposed lake bottom sediments through a variety of wind-abatement techniques pioneered at Owen's Lake (*11*).

Timothy J. Bradley* and Gregory M. Yanega
Department of Ecology and Evolutionary Biology, University of California, Irvine, CA 92697, USA.
*Corresponding author. Email: tbradley@uci.edu

REFERENCES

1. J. Fasberg et al., *Ecol. Monograph.* **80**, 3 (2010).
2. W. D. Shuford, N. Warnock, K. C. Molina, K. K. Sturm, *Hydrobiologia* **437**, 255 (2002).
3. W. D. Shuford, "Patterns of distribution and abundance of breeding colonial waterbirds in the interior of California, 2009–2012. A report of Point Blue Conservation Science to California Department of Fish and Wildlife and U.S. Fish and Wildlife Service (Region 8)" (2014); www.prbo.org/refs/files/12321_Shuford2014.pdf.
4. C. B. Wilsey, L. Taylor, N. Michel, K. Stockdale, "Water and birds in the arid West: Habitats in decline" (National Audubon Society, 2017).
5. E. P. Glenn, K. W. Flessa, J. Pitt, *Ecol. Engineer.* **51**, 1 (2013).
6. M. J. Cohen, "Hazards toll: The cost of inaction at the Salton Sea" (Pacific Institute, 2014); http://pacinst.org/publication/hazards-toll/.
7. Little Hoover Commission, "Averting disaster: Action now for the Salton Sea" (2014); www.lhc.ca.gov/report/averting-disaster-action-now-salton-sea.
8. N. Warnock, S. M. Haig, L. W. Oring, *Condor* **100**, 589 (1998).
9. A. L. Frie, J. H. Dingle, S. C. Ying, R. Bahreini, *Environ. Sci. Technol.* **51**, 8283 (2017).
10. Salton Sea Management Program, "Phase I: 10-year plan" (2017); https://resources.ca.gov/docs/salton_sea/ssmp-10-year-plan/SSMP-Phase-I-10-YR-Plan-with-appendices.pdf.
11. Great Basin Unified Air Pollution Control District, "Owens Lake dust control status" (2017); www.gbuapcd.org/owenslake/Landsat/dustcontrolstatuscurrent.htm.

10.1126/science.aar6088

Fund the Biological Survey Unit

The Biological Survey Unit (BSU) is administered by the U.S. Geological Survey (USGS) and located in the Smithsonian Institution's National Museum of Natural History (NMNH) (*1, 2*). Since 1889, the BSU has managed the

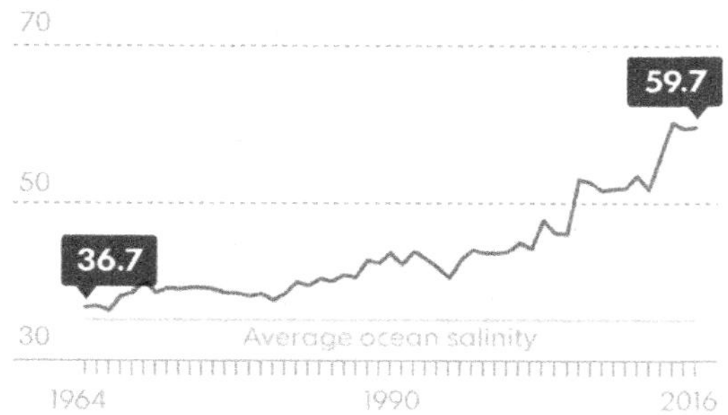

Environmental justice: California Air Resources Board warns dust can cause asthma, lung, heart disease, premature death placing elderly and children at greatest risk; Imperial County has lowest per capita annual income in California counties, highest unemployment rate (25%–2019 data), *highest* rates of childhood asthma.

Depletion — Groundwater in California's Central Valley

- Groundwater pumping has led to subsidence and collapse of parts of aquifer system.
- Solution? Reduce pumping; allow groundwater basins to recharge.
- US Geological Survey says if pumping ceased, it would take 50 years for Valley's aquifers to refill, as rain and snowmelt slowly seep underground.

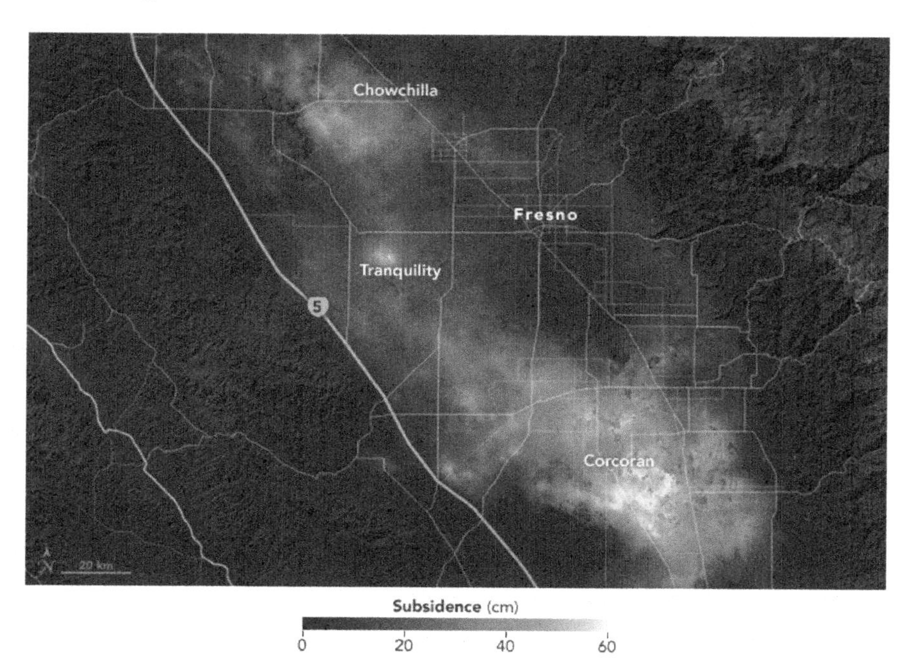

Total subsidence in the San Joaquin Valley May 7, 2015 — September 10, 2016 as measured by NASA and processed at JPL. Two large subsidence bowls are evident centered on Corcoran and El Nido with a small, new feature between them, near Tranquility.

Dwindling Groudwater — California's Central Valley

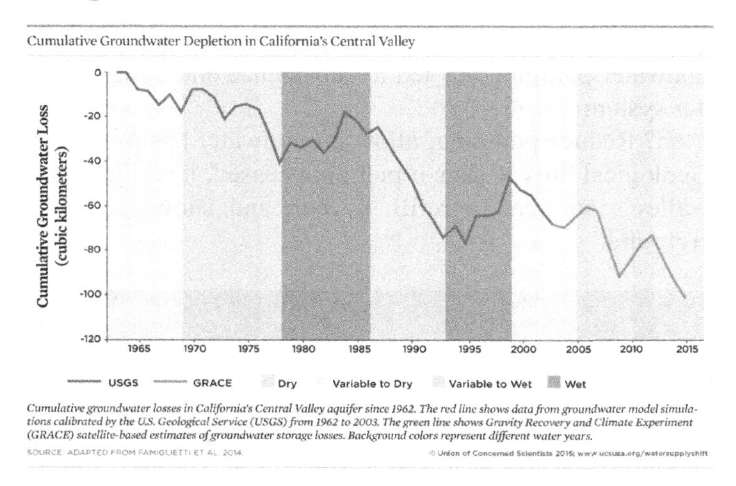

Cumulative groundwater losses in California's Central Valley aquifer since 1962. The red line shows data from groundwater model simulations calibrated by the U.S. Geological Service (USGS) from 1962 to 2003. The green line shows Gravity Recovery and Climate Experiment (GRACE) satellite-based estimates of groundwater storage losses. Background colors represent different water years.

SOURCE: ADAPTED FROM FAMIGLIETTI ET AL. 2014. © Union of Concerned Scientists 2015; www.ucsusa.org/watersupplyshift

San Joaquin and Central Valleys — Groundwater Depletion Impacts

Farm near Mendota, CA. Orchard near Coalinga, CA.

Over-pumping is pulling vast, unsustainable amounts of groundwater from the heart of California farming, the Central Valley, a UC Irvine satellite study shows — with enough lost over 4 years to fill two-thirds of Lake Mead — *2011 report.*

Groundwater Stewardship — It can be Done

- 2014 — *California Sustainable Groundwater Management Act* requires local governments to form groundwater sustainability agencies (GSAs).
- GSAs in high and medium priority basins must adopt groundwater sustainability plans by 2022.
- Once approved, GSAs have 20 years to implement plans and achieve sustainability.

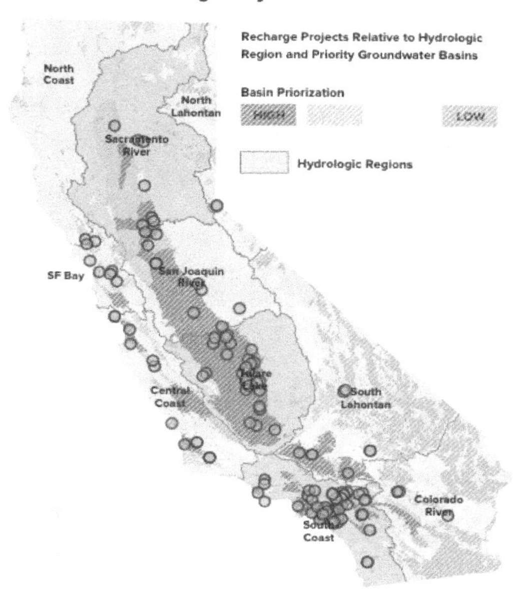

For Discussion

- Have we made progress in addressing diversion and depletion? Why, why not?
- Are their underlying causes being adequately addressed?

Degradation

Cities and their Rivers — Why do we Mistreat them?

Before and After — From Degradation to Restoration

Los Angeles Really has a River

- Until early 20th century — river was principal water supply for farms, orchards — via "zanjas" (ditches) that diverted water.
- 1939: transformed to a flood control corridor to permit commercial and residential development.

Los Angeles River — Modern History

Near Griffith Park (c. 1900).

Seasonal flooding (1941).

Flooding — Arroyo Seco (1913).

Walking the stream bed (1912).

Transforming the River — Army Corps of Engineers

"Building" the modern Los Angeles river (1938).

2019 — the river after it rains.

A River No More — Some Challenges

- Over 50 miles is channelized, straightened, encased in concrete — *it can't meander, percolate, or replenish groundwater.*
- A flood control channel, it receives contaminated storm-water runoff from streets and the built environment — *it's a pollution source.*
- With little native riparian corridor, it can't filter contaminants, provide nutrients or oxygen — *it supports little aquatic life.*
- Not an amenity — *provides few recreation, sightseeing, cultural opportunities.*
- An economic liability — *not an asset to attract jobs, housing, business investments.*

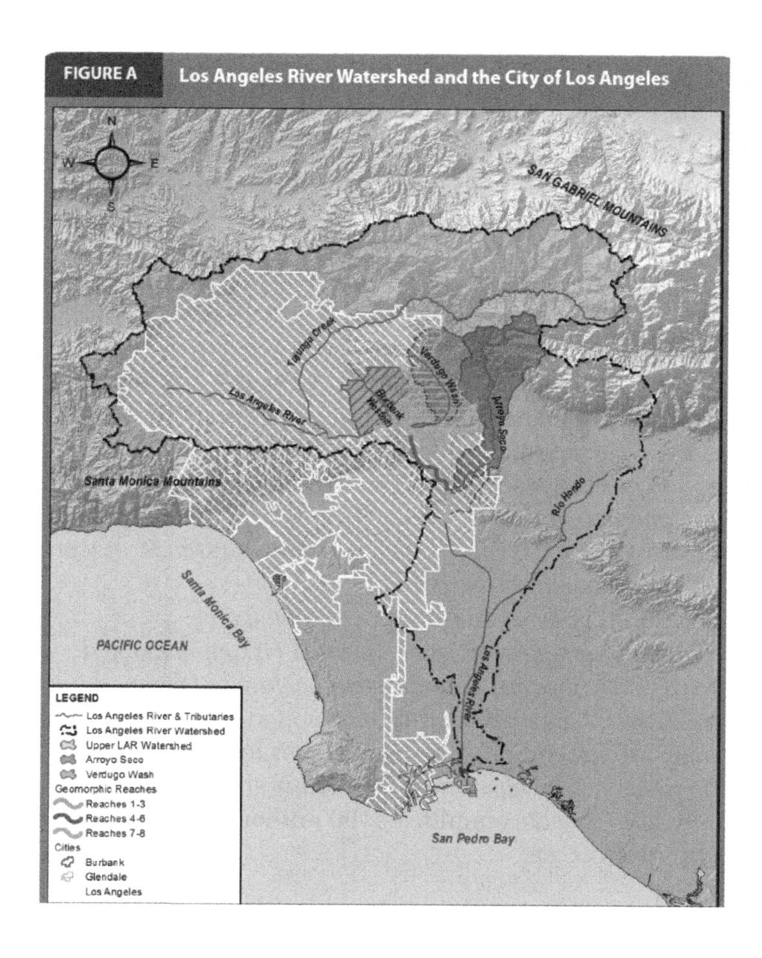

FIGURE A Los Angeles River Watershed and the City of Los Angeles

January 17, 2019 — Los Angeles River cresting toward "bank-full" condition.

Can it be Restored? The Los Angeles River *Revitalization Master Plan*

- Coalition: LA, LADWP, Friends of LA River — mission?
  - Remove channelization from Canoga Park to Vernon and replant native vegetation, develop greenways and parks, to:
    - *Improve water quality and ecosystem functioning (a healthy and active river).*
    - *Expand city's open space system/parklands via greenways and bikeways (recreation).*
    - *Create jobs — construction/maintenance.*
    - *Maintain existing levels of flood risk management.*
    - *Foster civic pride in cultural heritage of river.*
  - In 2014: Obama Administration supported $1 billion fund for 11 miles of restoration, from downtown to Elysian Park. Corps of Engineers, US Department of Interior also supported it.
  - July 2017: State committed $100 million for greenways.
  - http://lariver.org/

What Restoration Could do — Recreational Amenities

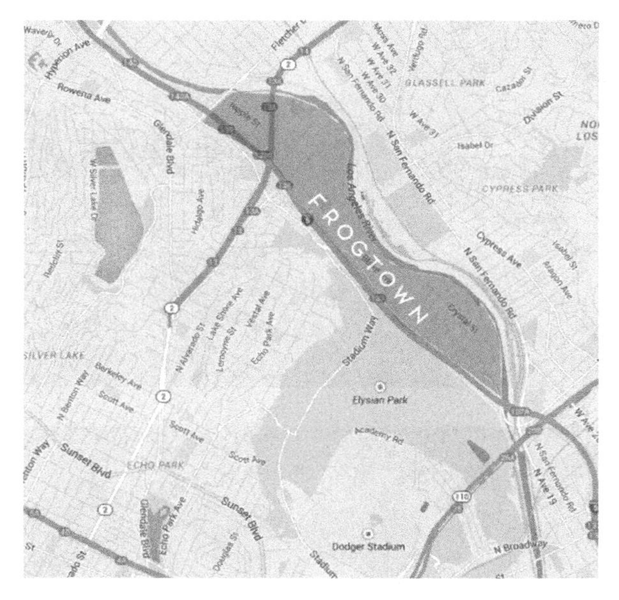

Who is Restoration for? Benefits and Costs

Golden Shore Marine Reserve, Long Beach — habitat restoration (2006).

Glendale Narrows Riverwalk (2012) — "In the past that's just been the back of Disney and Dreamworks lots. There's been graffiti, vandalism and gang activity back there, (now) character has been restored…".

- *Restoration or gentrification*? An *LA River Revitalization corporation* has been formed to guide riparian corridor development:
  - World War II era warehouses and other properties are becoming lofts and design studios.
  - "Pocket parks" are springing up in Maywood, South Gate, Long Beach, elsewhere.
  - But — rents are sharply rising in river-adjacent communities.
  - Environmentalists claim developers interested in profit dominate restoration plans.
  - Fears that older, lower-income families will be displaced are growing.

Restoration and Controversy

"Frank Gehry's controversial LA River Plan gets cautious, low-key rollout", Los Angeles Times, June 18, 2016.

HOW BROOKLYN GOT GENTRIFIED | **SEX CRIMES & REFUGEES**

A *NATION*/WNYC PODCAST — LAILA LALAMI

DOUBLE ISSUE

THE Nation. 150

LA *Lost, & Found*

Is it possible to beautify a city -- without making it a playground for the elite?

by RICHARD KREITNER

By Christopher Hawthorne, Architecture Critic

JUNE 18, 2016, 9:31 AM

The design team working with architect Frank Gehry on a controversial new master plan for the Los Angeles River has begun to introduce its work to the public — but in a noticeably cautious and low-key way.

River LA, the nonprofit group that began collaborating with Gehry's office more than a year ago, isn't ready to unveil any design proposals by the architect. Or any rough sketches, for that matter.

Instead it has been holding upbeat, informal "listening sessions" in neighborhoods near the river, in an apparent effort to build goodwill. The new website it developed with two of Gehry's partners, expected to go live Tuesday morning at lariverindex.org, is stuffed with maps and charts but similarly short on architectural detail.

- Focus of Gehry's work will be to help develop a new master plan.
- No specific designs have been unveiled and details remain scarce.
- *River LA* — the group hiring Gehry, is composed of real estate developers and land use attorneys, generating fears as to its goals.

River Degradation and Restoration are Global Challenges

Cheonggyecheon River, Seoul, Korea (clockwise from top):

- Pre-1945
- 1950s–2000s
- Today (2020)

How the Cheonggyecheon River was Restored

- By the mid-1950s, the Cheonggyecheon was so polluted, city officials decided that it couldn't be salvaged; it was paved over, completely hiding it from view, with a four-lane expressway.
- In 2001 former CEO of Hyundai Engineering and Construction, Lee Myung-bak, was elected mayor; promised to remove the freeway, restore the river.
- At the end of his term (2005), the Cheonggyecheon was flowing freely through the heart of downtown Seoul. The freeway was gone and a rapid transit bus line had taken its place.
- By 2006, hundreds of thousands of visitors had flocked to the Cheonggyecheon riverbanks; carp could be seen swimming in its pools. Myung-bak was elected president of South Korea in 2007.

Conclusion — Why do we Degrade Rivers and then Restore them?

- Changing urban needs and values influence how we view rivers, as well as our expectations regarding how a river should "function".
    - *Initial settlement* — exploit and use (emphasis on withdrawal and diversion).
    - *Industrialization* — harness and control (engage in channelization and damming).
    - *Post-industrialization* — restore and renew (reclaim and "naturalize").
- Post industrialization encourages people to value and seek natural amenities; thus, rivers are again viewed as:
    - Recreational assets.
    - Aesthetic assets.
    - Community development assets.
    - Social justice assets.
- *Urban planners* believe the public values of rivers and other waterways is becoming hotly contested — should they be managed as civic entities, gentrified landscapes, or both?

Section 7

California as "Hydraulic Empire" — Fact, Fiction, Fantasy?

California as "Hydraulic Empire" — Fact, Fiction, Fantasy?

- California faces a number of threats to its water management.
- Not all threats revolve around *drought or pollution* — some entail having too MUCH water in the wrong place and at the wrong time (i.e., flood).
    - Can we manage flood threats?
    - How doe our *built environment* worsen the threat of flood?
    - What other challenges do we face that need resolution?

Flood as Water Resource Policy Challenge

San Jose Mercury News: March 4, 2017.

More than a decade ago, the US Army Corps of Engineers considered a $7.4 million project that would have protected the Rock Springs neighborhood from last month's devastating floods. It concluded the project was too costly and refused to fund it.

The Corps worked for five years with the Santa Clara Valley Water District to design flood walls, levees and other improvements for the San Jose neighborhood, but in the end, Corps officials said they had no way to proceed because of a harsh calculus: *They are not allowed by federal law to build projects whose cost exceeds whatever damage would occur in a major flood.* And the projected damage from a flood in that area was relatively minor, the Corps concluded.

Flood and Flood Protection

- Flooding is partly a *natural* hazard, partly a human-made one.
  - Altering streams and floodplains by dredging, channelizing, stream straightening, and shoreline development increases *risk of property damage and loss of life*.
- Historically, we've tried to fortify ourselves against flood through dams and levees (purpose = protection from loss).
  - This creates *illusion of security* by inducing high-risk activities along floodplains.
  - *Encourages more building* within floodplains, exposing more property to damage, and generating the perception that we are safe.
- Authority for flood protection/alleviation is shared by different government levels — little coordination (power and process are shared and diffuse).

A Growing Threat — Sea Level Rise and Coastal Flooding

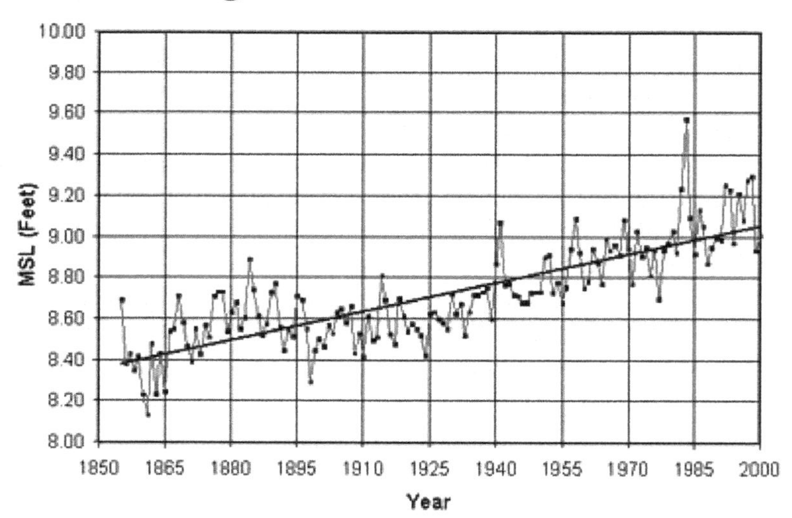

Source: California EPA — Golden Gate tidal gauge at entrance to San Francisco Bay.

- Sea level rise *will* affect inland areas as well as coastal zones (e.g., Sacramento/San Joaquin Delta levees) as we will see.

Sea Level Rise from Climate Change a Global Threat

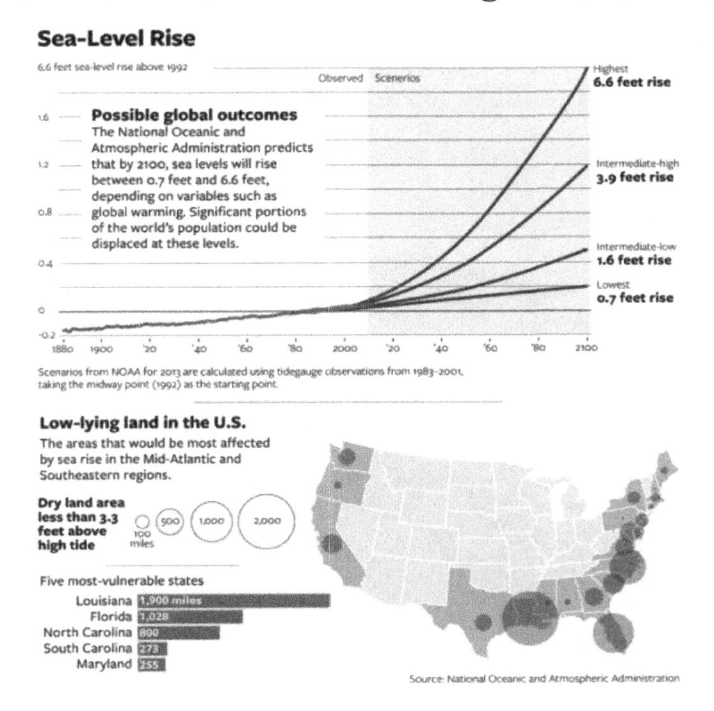

The Built Environment and Flood Risk

- Natural disasters?
- Human disasters?
- Both?

Newport Beach, 7/03/2020.

Newport Beach, 7/04/2020.

Santa Barbara County, 01/2019.

Solutions — Geo-Engineering?

1940

1997

2017

Portion of breeched levee — Sacramento River (1997).

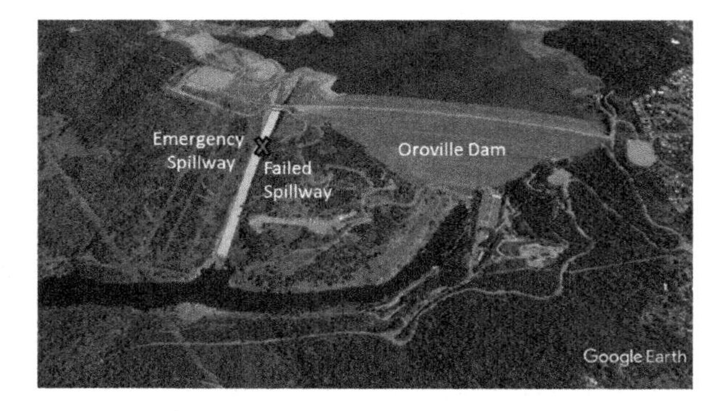

May 2015.

February 12, 2017.

Geo-Engineering Limits — California as Example

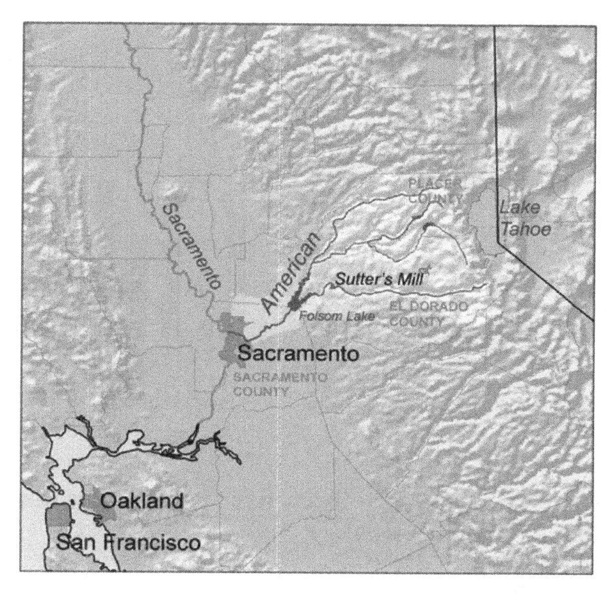

- Sacramento's risk of flooding is the greatest of any major US city. Why?
  - Water flowing out of Sierra Nevada during floods is increasing.
  - Folsom Dam was designed to reduce flood flows in American River — larger storms, more runoff have made the dam inadequate.
  - Core of today's levees built by settlers 150 years ago.
  - Early levees not constructed to current standards, little care given to suitability of foundation soils.

Historic Flood Flows — American River, Sacramento

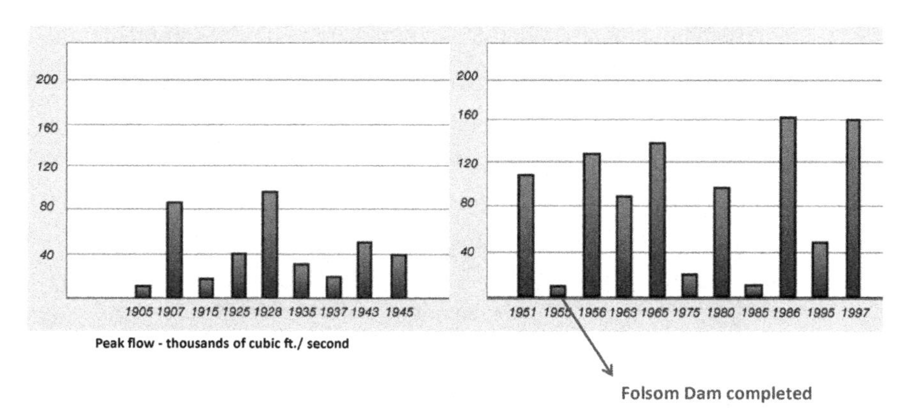

Since the late 1940s, there have been five floods larger than any previously recorded maximum floods.

Sacramento Area Flood Control Authority (SAFCA)

- 980 miles of levees, plus pumping plants and bypass channels.
- Designed to protect communities and farms of Sacramento Valley and Delta.

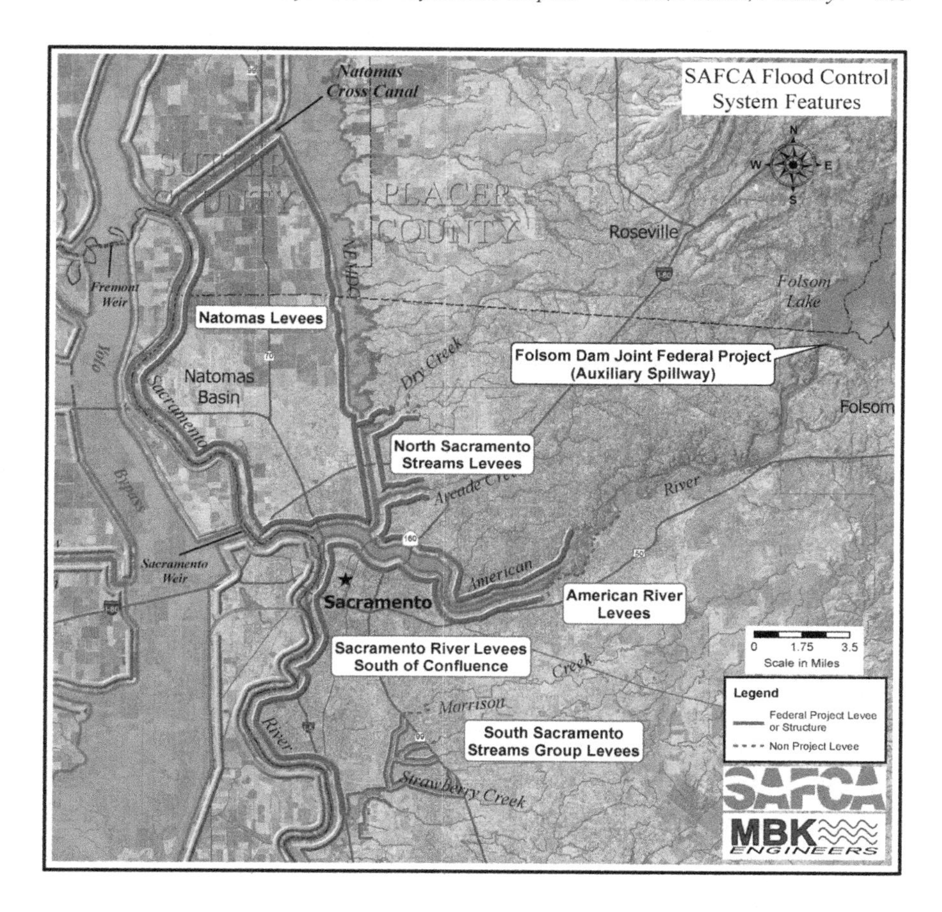

For Discussion

- Are engineering solutions the best way to avoid flood threats?
- What alternatives could you imagine?

What are Some Flood Risk Alternatives?

- Adaptive alternatives:
  - Avoid flood risks and employing "replaceable" or low-loss land uses.

- Adopt low-impact development alternatives in built-environments.
- Use major flood disasters to teach resiliency and how to avoid the worst possible hazards (Wilkman, 2017).

- Changing social perceptions and attitudes about flood:
  - Requires concerted effort, changes in incentives to avert risky behavior.
  - Farming is valuable, especially on lands prone to floods.
  - Levee systems protect *water supply and ecological resources but are limited in their effectiveness — must consider these limits.*

Why Geo-Engineering Alone Cannot Work — Sacramento's Levees

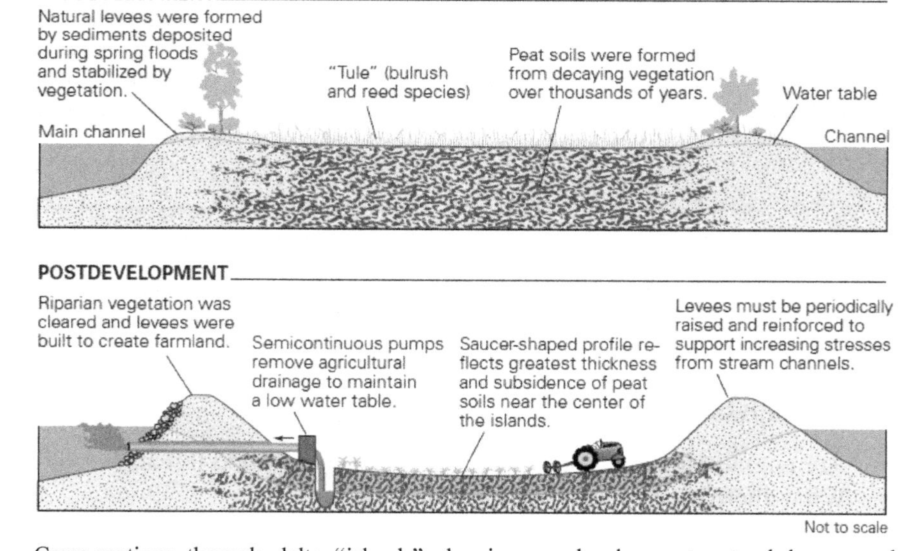

Cross-sections through delta "islands" showing pre-development natural levees and post-development constructed levees. *Images by USGS.

Sacramento's Remedies — Many Measures, Gradual Risk Reduction

- Since 2005, region has adopted several measures; cost = $4.1 billion:
    - State adopted land use and environmental enhancement policies for region (e.g., more wetlands restoration) to offset flooding — 2007.
    - Fortify existing levees on American River (but NOT building new dams) — 2008.
    - Offer *preferred risk flood insurance* to residents of floodplains — 2008.
    - Residents in "200 year floodplain" voted to assess special tax to finance local improvements.
    - SAFCA adopted a *Development Impact Fee* program to offset effects of future floodplain development and pay for flood system improvements — 2008.

…And Today?

UTILITIES HOME
PAY YOUR UTILITY BILL
WATER
DRAINAGE
STORMWATER
FLOOD READY
YOUR FLOOD PREPARATION
HOW TO PREPARE YOUR KIDS

FLOOD ZONE CONSTRUCTION REQUIREMENTS

The City of Sacramento has adopted building codes and procedures designed to protect lives and property in the event of a 100-year flood. Specific floodplain management regulations and building codes are enforced to regulate construction in at-risk areas throughout the city.

200-year floodplain requirements will be in effect July 2, 2016.

KNOW THE BASICS

- Get the proper permit before you start construction
- Don't dig, plant or build at the base of a levee
- Building near the levee? Know encroachment levee regulations. Visit Central Valley Flood Protection Board.
- Don't throw your trash or debris in streams, channels or open bodies of water
- Do not forget to elevate your HAVAC exterior units

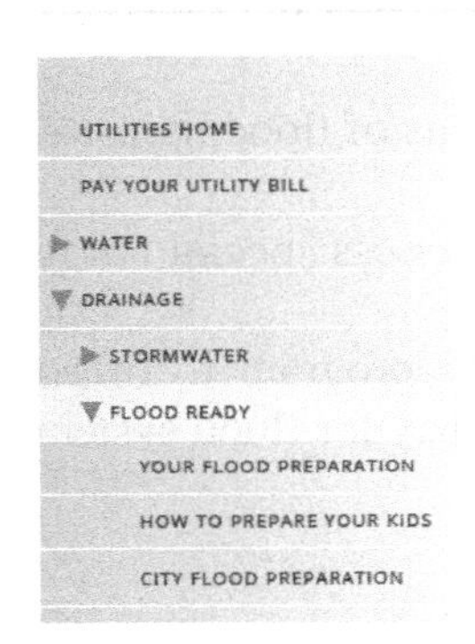

Home 311 Customer Service Emergency Management Contact Us How Do I... Skip to Content Search

City of SACRAMENTO

UTILITIES

BUSINESS | CITY HALL | LIVING HERE | ONLINE SERVICES | VISITORS

Home > Utilities > Drainage > Flood Ready

UTILITIES HOME
PAY YOUR UTILITY BILL
WATER
DRAINAGE
STORMWATER
FLOOD READY
YOUR FLOOD PREPARATION
HOW TO PREPARE YOUR KIDS
CITY FLOOD PREPARATION

FLOOD READY

The City of Sacramento strives to build a flood resistant and resilient community through preparedness and mitigation. Materials have been developed to provide residents and business owners with information on local flood risks and preparedness tools.

2021 LOCAL HAZARD MITIGATION PLAN UPDATE PROJECT

The City of Sacramento is partnering with Sacramento County, other incorporated communities, and numerous special districts to update their countywide 2016 Local Hazard Mitigation Plan (LHMP). Flood, drought, earthquake, and severe weather are just a few of the hazards to Sacramento communities. While natural hazards such as these cannot be prevented, an LHMP forms the foundation for a community's long-term strategy to reduce disaster losses by breaking the repeated cycle of disaster damage and reconstruction. Communities with a FEMA-approved LHMP are eligible for FEMA pre- and post-disaster grant funding and for lower costs of flood insurance to residents through the National Flood Insurance Program's (NFIP) Community Rating System (CRS). For more information on this project and how to get involved, go to www.StormReady.org.

https://www.cityofsacramento.org/Utilities/Drainage/Flood-Ready

Adaptive Solutions — Mapping of Risk, Educating about Risk

Environment and Behavior
1–29

Reprints and permissions:
sagepub.com/journalsPermissions.nav
DOI: 10.1177/0013916517748711
journals.sagepub.com/home/eab
SAGE

The Influence of Hazard Maps and Trust of Flood Controls on Coastal Flood Spatial Awareness and Risk Perception

Douglas Houston[1], Wing Cheung[2], Victoria Basolo[1], David Feldman[1], Richard Matthew[1], Brett F. Sanders[1], Beth Karlin[3], Jochen E. Schubert[1], Kristen A. Goodrich[1], Santina Contreras[4], and Adam Luke[1]

- Respondents who previously experienced flooding and who viewed the FloodRISE map, reported that they now had higher levels of flood hazard awareness — including role of sea level rise as cause.
- This suggests that provision of a hazard map to coastal residents in an interactive digital format that allows panning and zooming raises the awareness of respondents to sea-level rise and flooding.
- Providing a map with flood depth estimates and more precise ways of differentiating risk appears to heighten local flood hazard awareness.

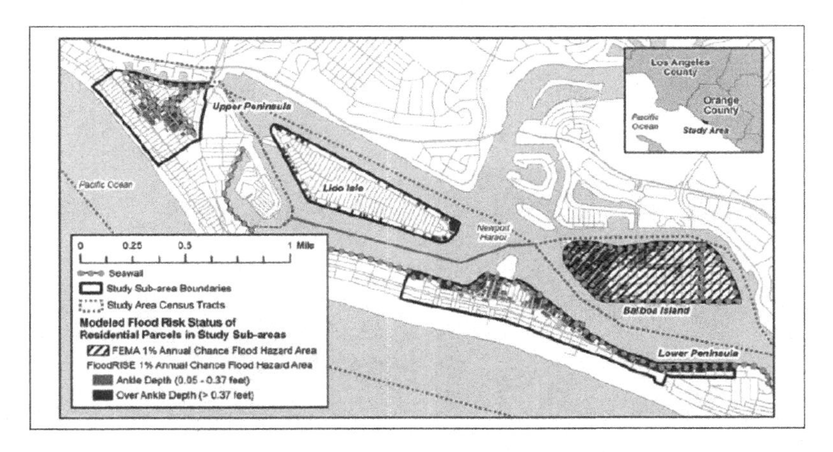

Figure 1. Study area boundaries and modeled flood hazard estimates.

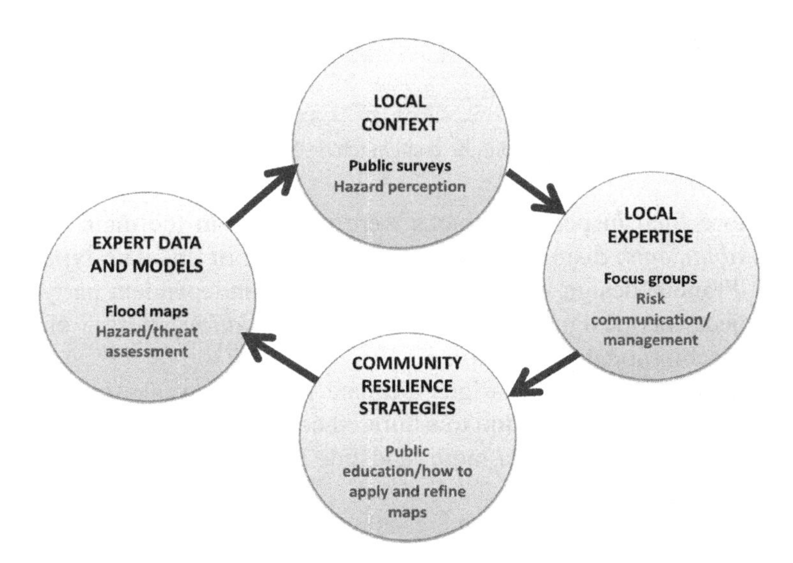

- Generate visualization tools (maps) depicting flood risk.
- Use surveys and focus groups to transform these tools into useful products that can help communities plan flood hazard avoidance, response, recovery.

And St. Francis Dam?

'All the News 'All the Time

Los Angeles Times

In Three Parts — 44 Pages

Liberty Under Law — Equal Rights — True Industrial Freedom

Vol. XLVII. WEDNESDAY MORNING, MARCH 14, 1928. DAILY, FIVE CENTS SUNDAY, TEN CENTS

200 DEAD, 30[illegible] MISSING, $7,000,000 LOSS IN ST. FRANCIS DAM DISASTER

Plane Views Give Idea of Magnitude of Catastrophe

Left—View of the wrecked St. Francis Dam taken from the air at a point just below the dam. Right—Santa Paula, as seen from the air. Directly in the path of the raging torrents, portions of the city were inundated. [Photographs taken by Harry C. Anderson, Times staff photographer.]

THOUSANDS RUSH TO AID IN WORK OF RESCUE AND RELIEF

Red Cross Directs Gigantic Task of Succor in Devastated Area; Water Board Appropriates $25,000; Food and Shelter Needed Immediately

PARTIAL LIST OF PERSONS DEAD, INJURED AND MISSING IN FLOOD

The following is a partial and best obtainable list of identified dead, unidentified dead, missing and injured in the St. Francis dam break yesterday.

As well as the below listed persons the authorities are searching for hundreds of missing men, women and children in San Francisquito Canyon, Fillmore, Saugus region, Piru, Ventura, Santa Paula, Castaic, Oberg, Camulos and other sections.

According to figures from improvised morgues in the stricken area and from the Sheriff's office and police department more than 200 bodies of men, women and children have been recovered of which in the [illegible]

SCORES MORE THOUGHT BURIED IN DEBRIS OF WILD WATERS

Darkness Halts Rescue as Crews Sight Other Bodies; Hundreds of Homes Crushed by Fatal Torrent and Many Bridges Washed Away

Approximately 200 persons dead, 300 others unaccounted for, and property damage estimated [illegible]

https://www.youtube.com/watch?v=YWf6H3l4T4E.

- Before final inspection, cracks were observed in the main dam and its *abutments*; dismissed as conditions typical of this dam type.
- St. Francis' design was not reviewed by any independent party.
- Project was designed to prevent small *foundation* stresses only and not accommodate full *uplift*.
- Post-disaster analysis assigns ultimate failure to weakening of abutment foundation rock due to saturated condition created by the reservoir. This re-activated a large landslide that caused failure to initiate at dam's left end.
- Entire dam tilted and rotated, causing catastrophic failure .
- *There were no adequate warning or evacuation systems in place.*

Applying Adaptive Solutions — Bangladesh

- 160 million residents — chronic flooding from cyclones, tropical storms kills thousands.
- Sea level rise may displace 15% of population.
- Ganges, Brahmaputra Rivers constantly shift.
- In 1990s, $10 billion *World Bank* project, backed by France, Japan, US, proposed 8,000 km of dikes to control rivers, hold back sea:
  - Farmers opposed project because their lands would be taken; UN concluded plans were impractical because local soils too unstable.

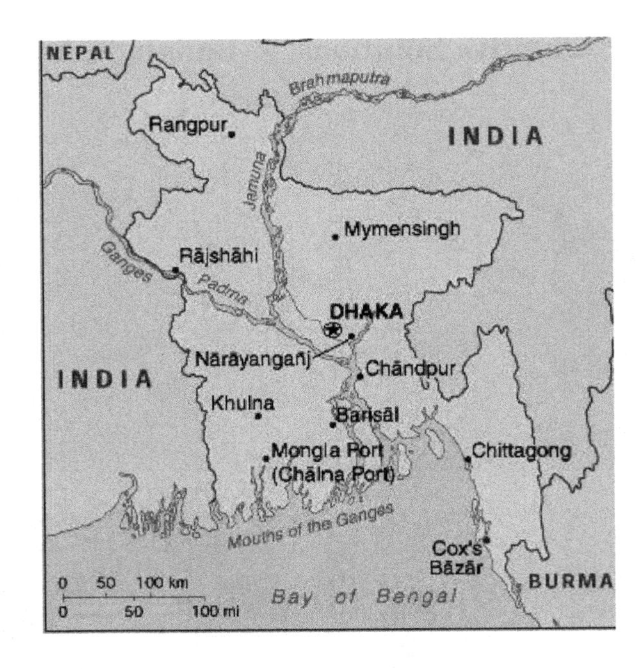
NEPAL
Brahmaputra
Rangpur
INDIA
Jamuna
Mymensingh
Rājshāhi
Ganges
Padma
DHAKA
Nārāyanganj
Chāndpur
INDIA
Khulna
Barisāl
Mongla Port
(Chālna Port)
Chittagong
Mouths of the Ganges
Cox's
Bāzār
0 50 100 km
0 50 100 mi
Bay of Bengal
BURMA

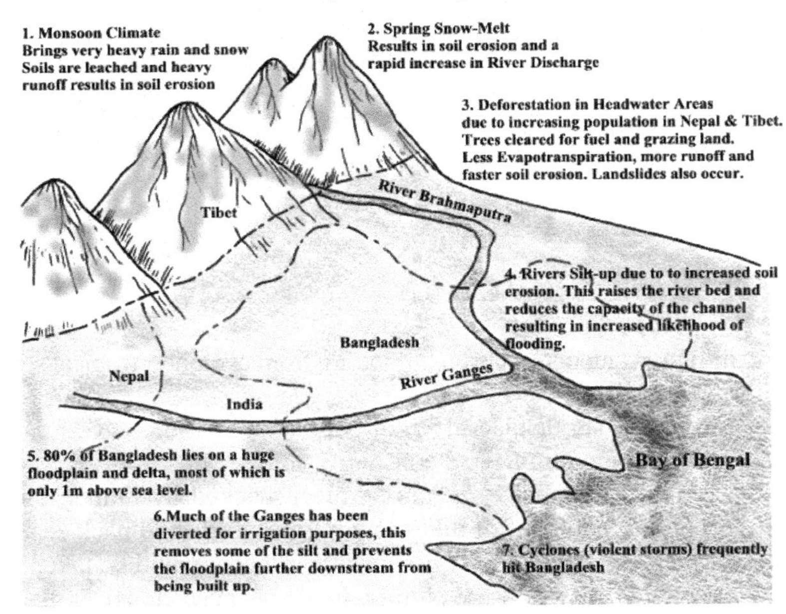
Some Causes of Flooding in Bangladesh
1. Monsoon Climate
Brings very heavy rain and snow
Soils are leached and heavy
runoff results in soil erosion
2. Spring Snow-Melt
Results in soil erosion and a
rapid increase in River Discharge
3. Deforestation in Headwater Areas
due to increasing population in Nepal & Tibet.
Trees cleared for fuel and grazing land.
Less Evapotranspiration, more runoff and
faster soil erosion. Landslides also occur.
River Brahmaputra
Tibet
4. Rivers Silt-up due to to increased soil
erosion. This raises the river bed and
reduces the capacity of the channel
resulting in increased likelihood of
flooding.
Bangladesh
Nepal
River Ganges
India
5. 80% of Bangladesh lies on a huge
floodplain and delta, most of which is
only 1m above sea level.
Bay of Bengal
6.Much of the Ganges has been
diverted for irrigation purposes, this
removes some of the silt and prevents
the floodplain further downstream from
being built up.
7. Cyclones (violent storms) frequently
hit Bangladesh

Adaptation Through Empowerment

- NGOs (Oxfam, *Practical Action*, *CARE*) and villagers have developed local-scale, low-tech measures:
  - Raised dwellings topped with inexpensive paneled homes — less likely to be washed away.
  - *Introduce novel farming techniques*; e.g., floating gardens; salt-tolerant rice; convert rice paddies to shrimp ponds.
- Lessons? Incorporating voices of those impacted by flood produces better innovations. Partly made possible by country's long-experience with crisis management.

Floating farm.

Floating gardens.

Stilt homes — Gazipara, Bangladesh.

Other Water Risks: Fracking

- Fracking: injecting water and chemicals at high pressure into shale rock formations to fracture rock/extract gas and oil:
  - Often uses toxic chemicals (composition of which is proprietary and not disclosed).
  - 2001 — 2% of US gas and oil were produced through hydraulic fracturing; today — 35%: USA is clearly the world leader; China NOT far behind.
  - Spreading to Europe (UK, Poland — especially), Asia, Australia.
- Highest risks are to water quality — improper storage and handling of fluids at well site; spills and improper lining of pits; injection of wastewater into disposal wells, which can trigger earthquakes; potential for groundwater contamination.

Fracking — A Schematic

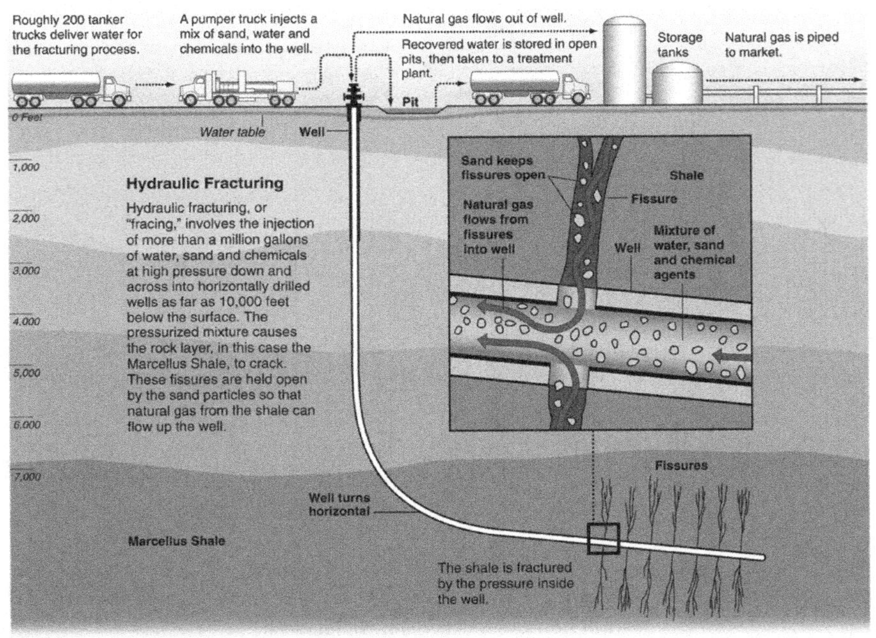

Graphic by Al Granberg

Fracking — Risks to Groundwater

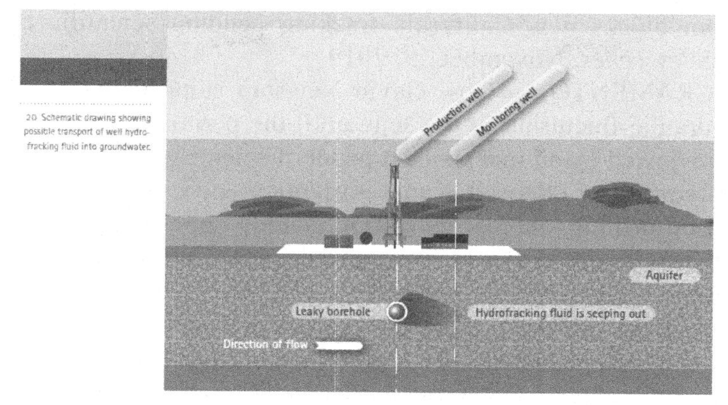

From: Hydrofracking risk assessment — Helmholtz environmental research center, Germany and ExxonMobil, 2012).

Adaptively Managing Fracking

- High uncertainty and lack of peer reviewed scientific studies on the potential risks to water quality.
- *Lack of credible information is not readily available to facilitate sound policy* (Wheeler Institute for water law and policy UC Berkeley), and no single model available for how to manage:
    - *European Union* (2012) — adopted a risk-based regulatory framework to provide "clear, predictable and coherent approach to unconventional fossil fuels ... development to allow optimal decisions to be made in an area where economics, finances, environment and in particular public trust are essential".
    - US (2013) Natural Resources Defense Council proposes putting critical watersheds off limits; requiring strongest well siting, casing and cementing and other drilling best practices; allowing communities to restrict fracking through zoning and planning — EPA has no national policy.
    - China (2014) — has world's largest shale-gas reserves, wants to produce 6.5 billion m^3 of shale gas/year; regions that would be most affected are severely water stressed (<2000 m^3 available per person) — would compete with agricultural, industrial, domestic sectors (*Science*, 2013).

Fracking — California

"Newsom blocks new California fracking pending scientific review", *Los Angeles Times*, November 19, 2019

SACRAMENTO — Gov. Gavin Newsom stopped the approval of new hydraulic fracturing in the state until the permits for those projects can be reviewed by an independent panel of scientists.

Newsom also imposed a moratorium on new permits for steam-injected oil drilling, another extraction method opposed by environmentalists that was linked to a massive petroleum spill in Kern County over the summer.

"These are necessary steps to strengthen oversight of oil and gas extraction as we phase out our dependence on fossil fuels and focus on clean energy sources", Newsom said in a statement Tuesday morning. "This transition cannot happen overnight; it must advance in a deliberate way to protect people, our environment, and our economy".

Along with halting use of the oil extraction methods, the Newsom administration plans to study the possible adoption of buffer zones around oil wells in or near residential neighborhoods, schools, hospitals and other facilities that could be exposed to hazardous fumes.

Gov. Gavin Newsom, left, is briefed by Billy Lacobie of Chevron, center, and Jason Marshall of the California Department of Conservation's Division of Oil, Gas and Geothermal Resources in July. A spill of more than 800,000 gallons flowed into a dry creek bed in an oil field west of Bakersfield.

Conclusions — Managing a Hydraulic Empire

- *Acknowledge* importance of broad-scale participation in decisions.
- *Admit* that no group has a monopoly of knowledge about water — *incorporating local knowledge in important.*
- *Emphasize* adaptive *solutions that are small in scale, incremental, reversible.*
- *Understand* that control over water must be tempered by *fairness and accountability: if encouraged to participate, people will innovate.*

Section 8

California's Water — Problems and Solutions in Global Context

Is this a Solution?

- Conceived by a South African, Nicholas Sloane, to solve Cape Town's chronic water shortage — June 2019.
- "Southern Ice Project" would cost >$200 million, financed by two South African banks and Water Vision AG, a Swiss water technology and infrastructure company.
- Requires an agreement with South Africa to buy the Antarctic water.
- In 1940s, John Isaacs of the Scripps Institution of Oceanography began exploring transporting an 8 billion-ton iceberg to San Diego to mitigate California droughts.
    - One Cape Town official says: "Such a project is both complex and risky with an anticipated very high water cost. The greatest challenges pertained to containment and transportation of the melt water as well as its injection into the water supply system".
    - Others say looking beyond traditional water sources is necessary: "We do not have the luxury to discard options", says Dhesigen Naidoo, CEO of South Africa's Water Research Commission, a nonprofit funded by the country's water tax.

California's Biggest Water Need — Resilience

- Droughts are natural — climate change makes them more frequent, intense.
- Our behavior worsens their effects.
- Ecological impacts are also significant.
- Resilience = transition to renewable, low-energy water sources, integrated supply and demand management. How?
  - Conserve and avoid waste.
  - Become more efficient.
  - Re-use every drop.
  - Harvest new sources.

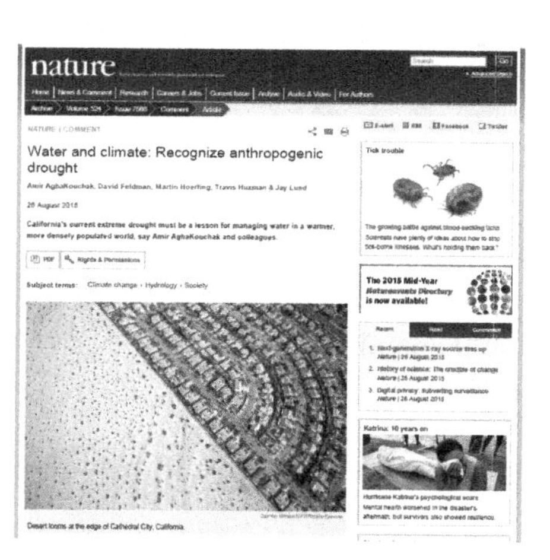

What Makes an Option Resilient?

- *Technical feasibility* — do science and engineering support its application?
- *Cost* — is it affordable relative to likely alternatives and who pays?
- *Environmental impact and risk* — can any adverse impacts be mitigated?
- *Public acceptability and engagement* — how do communities perceive the option; do they have a voice in its implementation?

Can we Build More Dams and Diversion Projects?

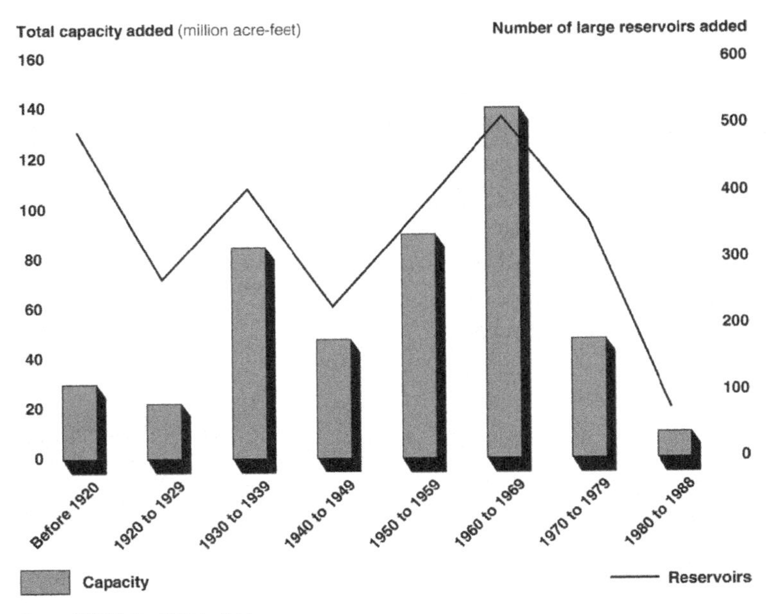

Number and size of reservoirs completed by decade.

- Few good sites remain (CA has 1,400 dams).
- High cost — states and users must be willing to cost-share.
- Environmental impacts high.
- Public acceptance low — in many cases, more support for *removing* dams than building new public works.

Can we Get More Water from the Delta?

- California's north-south water delivery system receives runoff from 40% of state's land area; accounts for 1/2 of total stream-flow.
  - State and Federal contracts provide for export of up to 7.5 million acre-feet per year through delta.
  - 83% of water goes to agriculture; 17% urban uses in central and southern California (27 million people rely upon it!)
- *Challenge: the freshwater used by cities and farms is also needed to support delta ecosystem.*

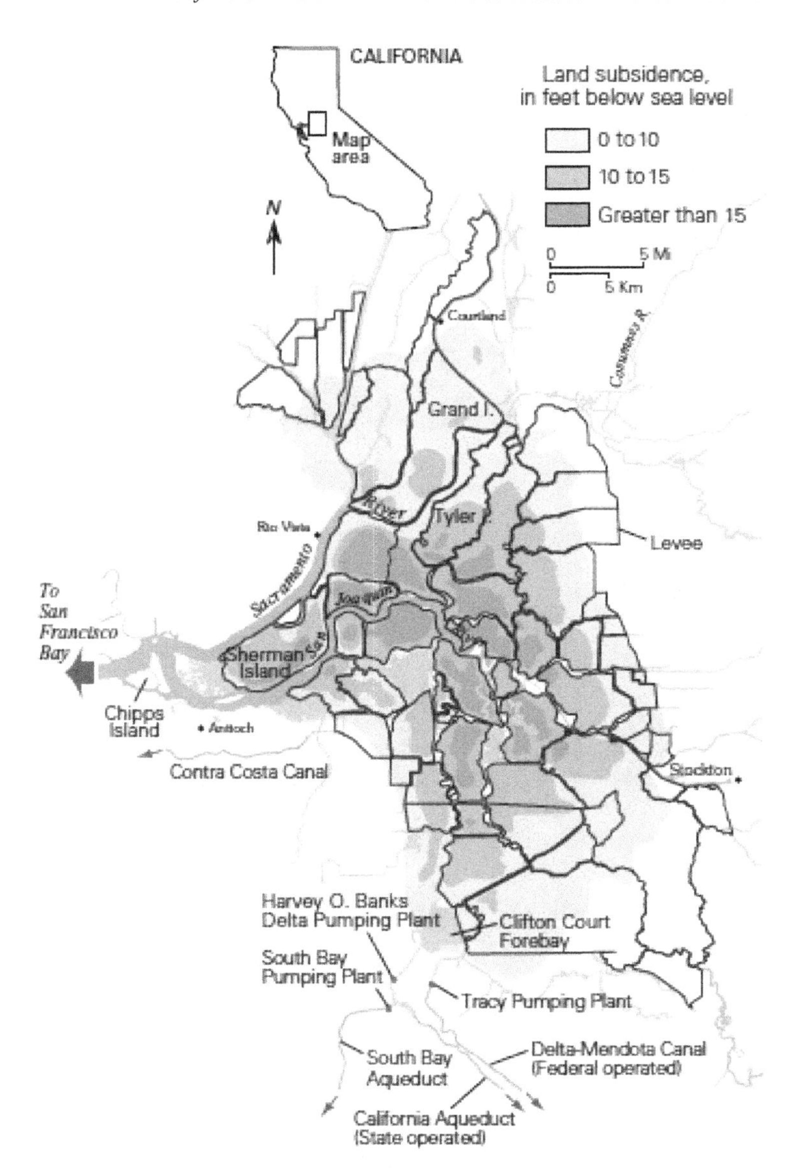
CALIFORNIA
Map area
Land subsidence, in feet below sea level
0 to 10
10 to 15
Greater than 15
0 5 Mi
0 5 Km
N
Courtland
Cosumnes R.
Grand I.
Tyler I.
Sacramento River
Rio Vista
Levee
San Joaquin
To San Francisco Bay
Sherman Island
Chipps Island
Antioch
Contra Costa Canal
Stockton
Harvey O. Banks Delta Pumping Plant
Clifton Court Forebay
South Bay Pumping Plant
Tracy Pumping Plant
South Bay Aqueduct
Delta-Mendota Canal (Federal operated)
California Aqueduct (State operated)

Can we Tunnel Under the Delta?

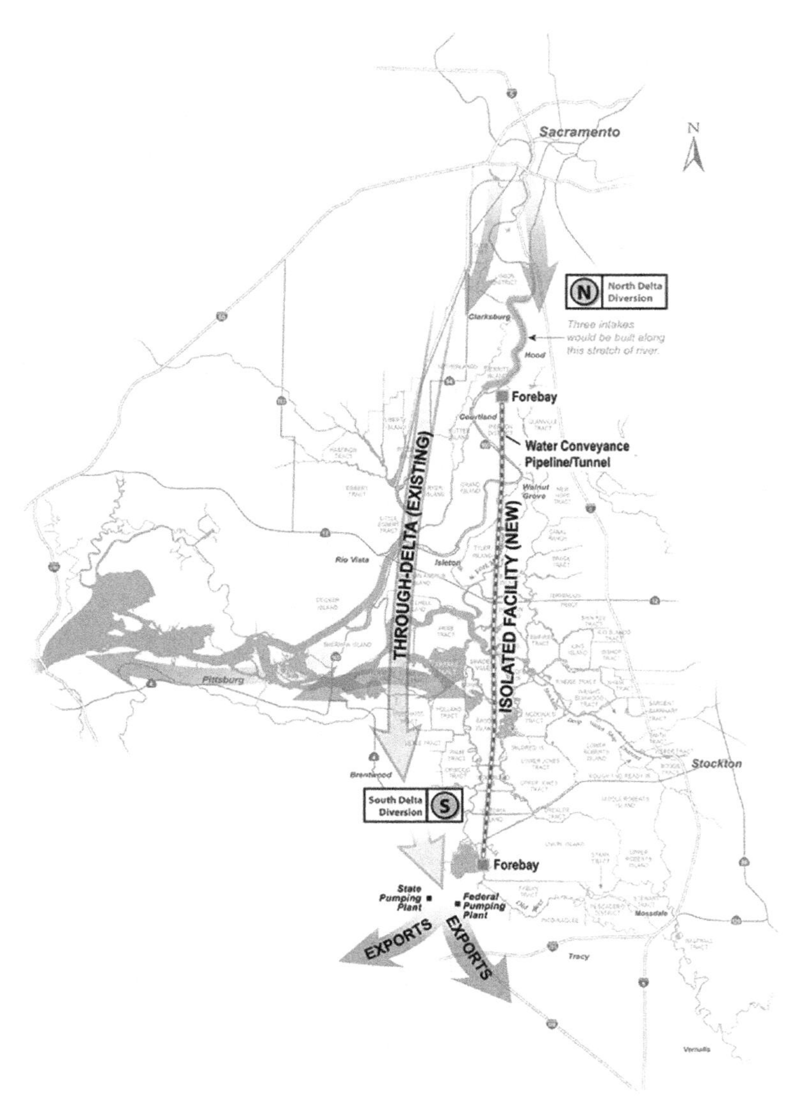

Delta Tunnels — Pro and Con

- Estimated cost = $17 billion — farmers, residential customers would pay for tunnels; taxpayers would fund habitat restoration.
- Advocates argue it will:
  - Reduce fish kills by replacing current delta pumps with gravity-flow tunnels.
- Protect freshwater diversions from levee collapse and sea level rise.
- Critics argue:
  - Will lead to higher water rates.
  - Other alternatives not sufficiently exhausted — e.g., "water neutral development".
  - *We need to fix the delta delivery system one way or another — levees are not stable.*
- *February 2019 — Gov. Newsom reduces project to ONE tunnel after farm interests balk at funding.*
  - https://www.vcstar.com/story/news/2019/02/12/newsom-downsizes-delta-water-project-one-tunnel-not-two/2855872002/.

Seawater Desalination?

Questions?

- Cost
- Energy consumption
- Entrainment and brine disposal
- Aesthetics
- Synergy with other sources*

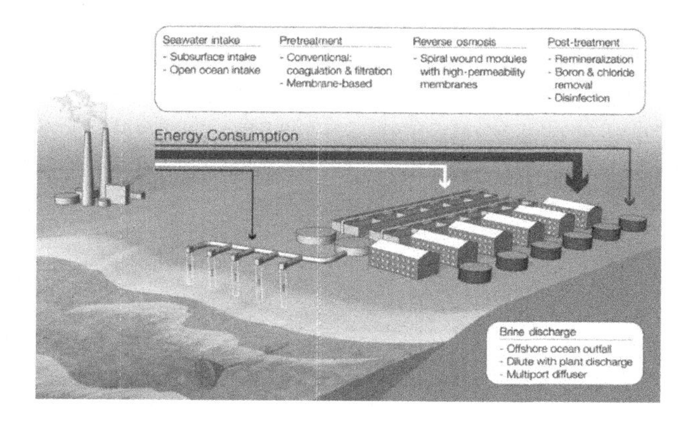

Feasibility and Economics — Energy Consumption

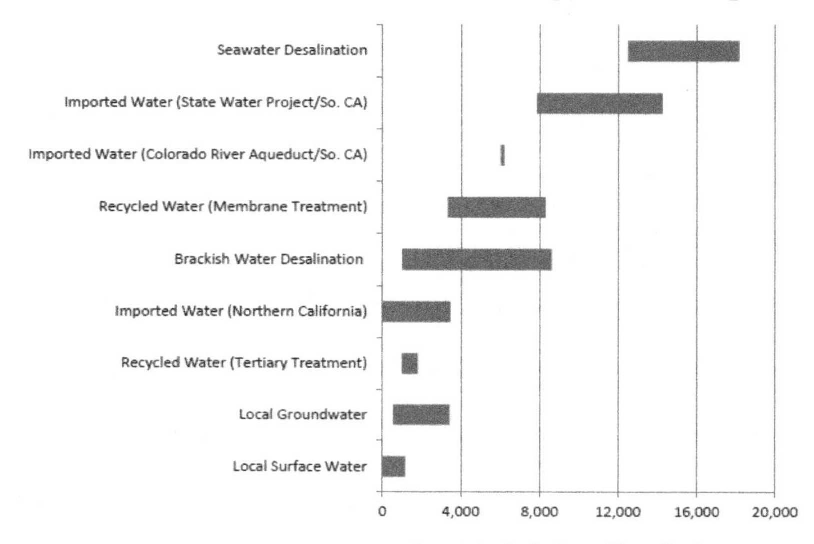

Figure 2. Comparison of the Energy Intensity of California Water Supplies

Notes: Estimates for local and imported water sources shown here do not include treatment, while those for desalination and recycled water include treatment. Typical treatment requires less than 500 kWh per million gallons. The upper range of imported water for Northern California is based on the energy requirements of the State Water Project along the South Bay Aqueduct. Energy requirements for recycled water refer to the energy required to bring the wastewater that would have been discharged to recycled water standards. Estimates for brackish water desalination are based on a salinity range of 600 - 7,000 mg/L.

Sources: Veerapaneni et al. 2011; GWI 2010; Cooley et al. 2012; GEI Consultants/Navigant Consulting, Inc. 2010

For Discussion

- What are some resilient water resource practices we should adopt?
- Why is resilience a difficult goal to achieve?

Addressing Concerns

Some answers?

- On-site energy provision
- Diffusers for brine
- Long-term rate agreements
- Integration with other sources*
- Public support*

CARLSBAD PROJECT (2015):

- Cost $950 million
- Water unit cost = $1800 acre-ft.
- 50 million gallons/day — serves 300,000

International Lessons — Israel

- After major drought (2000–2006), built 5 desal plants that provide 80% of urban supply: NOT a panacea — *Israel Water Authority* contends climate change will worsen future water availability.
- Country also re-uses wastewater for agriculture, uses drip irrigation, water-saving appliances.
- Broad consensus over water as a security issue induces cooperation with neighbors (e.g., Red — Dead desalination/basin transfer project with Jordan).

Israel allocates withdrawal permits — allocation not “rights” based.

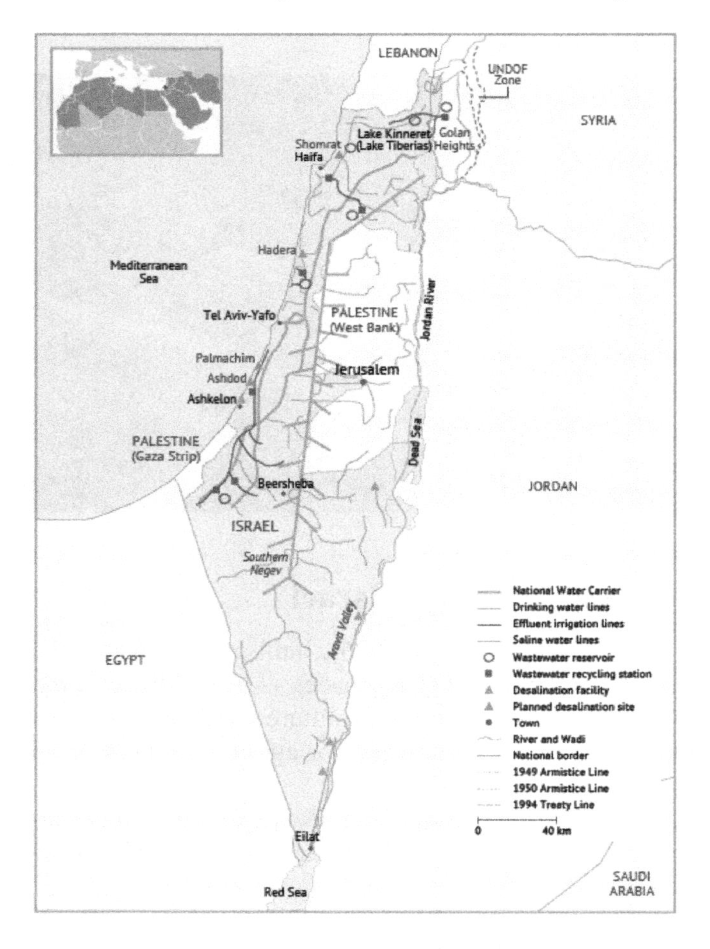

How About Re-Using Wastewater?

Concerns

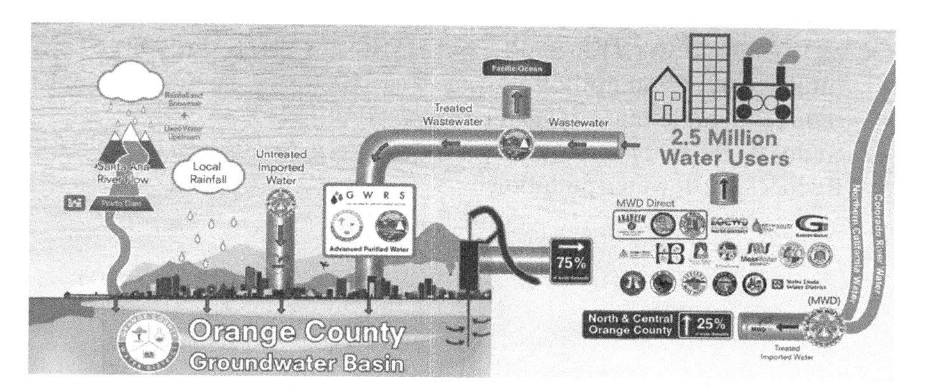

- Cost
- Energy consumption
- Health impacts
- Integration with other sources*
- Stigma, other concerns

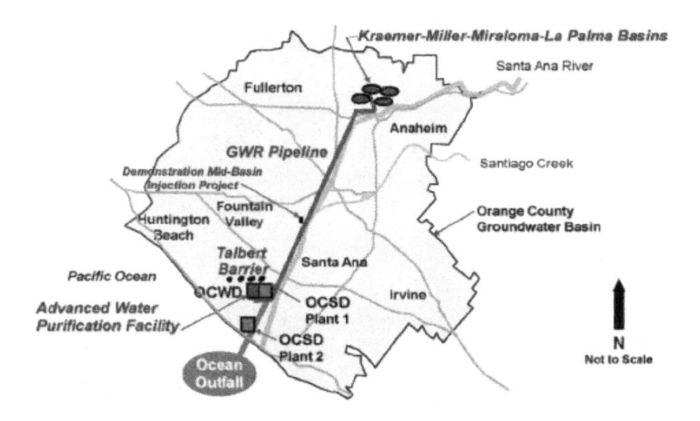

Addressing Concerns

Some answers:

- Built on-time, within budget
- Integration with other sources
- Employs advanced treatment*
- Ensures continued growth

OCWD GROUNDWATER REPLENISHMENT SYSTEM (2008):

- Initial cost = $750 million.
- Indirect potable reuse system (>100 MGD)
- Reduces needs for imported freshwater
- Reduces wastewater-pollution

Stages of Wastewater Treatment

Primary	Removal of a portion of the suspended solids and organic matter form the wastewater.
Secondary	Biological treatment to remove biodegradable organic matter and suspended solids. Disinfection is typically, but not universally, included in secondary treatment.
Advanced treatment	Nutrient removal, filtration, disinfection, further removal of biodegradable organics and suspended solids, removal of dissolved solids and/or trace constituents as required for specific water reuse applications.

SOURCE: Adapted from Asano et al. (2007).

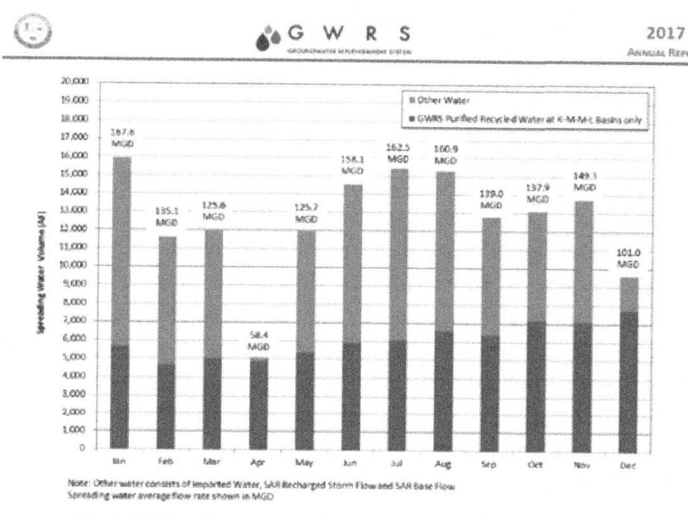

Figure 5-1. 2017 Spreading Water Sources and Volumes in the Anaheim Lake/Mini-Anaheim Lake/K-M-M-L/La Jolla Basins

Public Confidence and Technology Acceptance

- In less affluent areas, those with environmental justice concerns, *reuse* arouses *mistrust, especially among under-represented groups*.
- To become accepted:
  - All feasible options assessed and compared.
  - Management agencies perceived as trustworthy and competent.
  - Operations transparent and continuously monitored.
  - Implementation viewed as equitable.
- The GWR employs concerted public outreach: programs that emphasize operations, safety, benefits (economically as well as environmentally), and stresses comparative advantages.

Los Angeles — Prospects for Re-Use

L.A.'s bold goal to turn waste to drinkable water *LA Times, February 23, 2019*

In a dramatic shift for a city notorious for looking afar for most of its water, Mayor Eric Garcetti vowed this week that the city will be recycling all of its wastewater by 2035 and using it to reduce its need for imported supplies. "It is really a game-changer", said Richard Harasick, senior assistant general manager at the LA Department of Water and Power. Currently, recycling provides only 2% of the city's water. The Garcetti administration says that figure can jump to 35% if Los Angeles stops dumping its treated effluent into the sea and instead uses it to replenish the local groundwater reserves that help supply municipal customers. Doing that will require costly equipment upgrades at the Hyperion Water Reclamation Plant, new groundwater wells, construction of a 15-mile pipeline and as much as $8 billion financed by DWP to pay for it all. The plan will also require a change of heart by LA residents, who 18 years ago succeeded in killing a city project that would have used treated sewage to recharge the San Fernando Valley aquifer. City officials are optimistic. They say years of drought, declining imports and the high profile of a similar program in Orange County have softened resistance. "People are accepting it now", Harasick said.

GENARO MOLINA Los Angeles Times

A KEY ELEMENT in L.A.'s water recycling plan is the Hyperion water plant, which the city wants to upgrade so purified water can be pumped inland and injected into the aquifers that underlie the L.A. Basin.

GENARO MOLINA Los Angeles Times

L.A. MAYOR Eric Garcetti tells a crowd at the Hyperion facility that the city will be recycling all of its wastewater by 2035 to reduce the need for imported supplies.

International Lessons — Australia

- During *millennium drought* (1996–2010) employed public engagement to encourage re-use mostly for non-potable use.

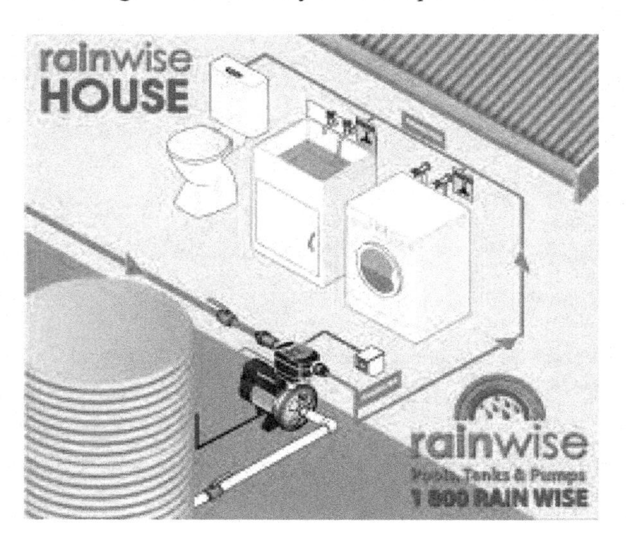

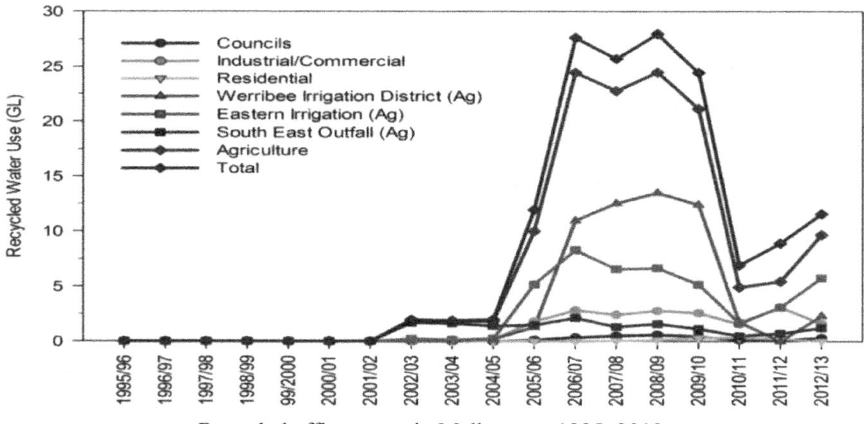

Recycled effluent use in Melbourne, 1995–2013.

Conservation — What Works?

- *Allocation-based rate structures*:
  - Tiered-rate systems reduce residential water use by 50% or more.
  - Utilities can apply a property-specific water budget to a household.

- *Water-smart billing systems*:
  - Showing residents how much water they use compared to "average" and "efficient" households — we want to be in line with their neighbors — we call this "norming."

- *Incentives to reduce outdoor uses*:
  - Rebates for converting lawns to drought-tolerant landscaping — *Las Vegas has reduced outdoor water use by 75%* — as well as high-efficiency landscape irrigation.

Water Conservation in Landscaping Act of 2006 imposes performance standards and labeling requirements for landscape irrigation equipment (e.g., irrigation controllers, moisture sensors, emission devices, valves) to reduce wasteful consumption.

...What Doesn't Work

"Shaming" or "outing" does not work — may have a cathartic effect, but evidence suggests it may have unintended consequences.

"The World is watching a lot more", says Tony Corcoran, one of several people who spend their spare time these days canvassing the communities of Beverly Hills, West Hollywood and elsewhere, looking for people wasting water during the worst California drought in recent memory. . . Not everyone is happy about it. One woman, quickly tiring of Corcoran's lecture on conservation while she watered her plants, turned her hose on him. — *Los Angeles Times, June 11, 2015.*

Resilience and the Environment or, "Fish Need Water Every Day"

1987 floods — normal riparian condition.

Mid-1990s — vegetation growth, same site

Colorado River mile 55 (s. of Glen Canyon Dam).

- Long-term recovery remains uncertain — e.g., endangered Chub fish — will habitat be sufficient?
- Drought, climate variability make full recovery unpredictable.
- 2005 report, Fish and Wildlife Service stated: "native fish recovery and sediment transport have not entirely improved".

Glen Canyon Dam Adaptive Management Program

- Begun (1992) to restore habitat and historic flows; involves seven states, two environmental groups, six native American tribes, several federal agencies.
- Interior Department (2012) approved long-term experimental program of high-flow releases and native fish protection to preserve and improve the Grand Canyon and its resources.
- GOAL: Change flows through *Glen Canyon Dam* to restore sand bars, riparian vegetation, fisheries, cultural artifacts — to resemble historic patterns.

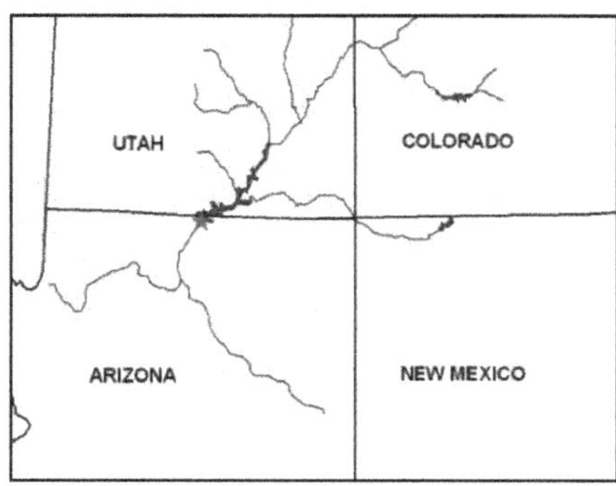

Other Examples — "Mal-Adaptive" Solutions

Endangered Species Act of 1973: "(V)arious species of fish, wildlife, and plants in the US have been rendered extinct as a consequence of economic growth and development . . . . (Other species) have been so depleted in numbers that they are in danger of or threatened with extinction. . . . (They are of) esthetic, ecological, educational, historical, recreational, and scientific value and (agencies must) use all methods and procedures . . . to bring (them) to the point at which the measures provided pursuant to this Act are no longer necessary".

Fish ladders: Bonneville Dam.

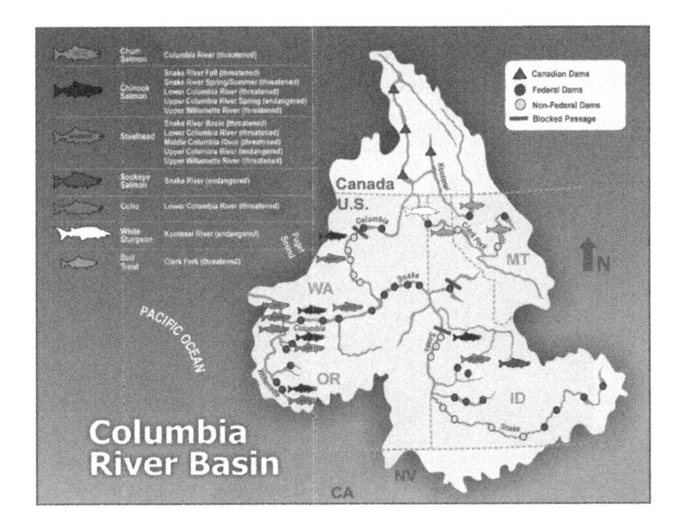

"Delta Smelt, Icon of California Water Wars, Is Almost Extinct — Tiny fish's survival hangs in the balance as severe drought and decades of water pumping drain its habitat". Jane Kay, National Geographic, April 3, 2015.

Conclusions

- *Climate challenges aren't new*: We've always been deniers of the impacts of climate (e.g., "rain will follow the plow").
- *Don't blame population growth: profligate consumption isn't caused by how many of us use water, but by HOW we use it.* Large scale, publicly-subsidized water projects are partly responsible for inducing profligacy.
- *There are no panaceas*: we large-scale infrastructure. We ALSO need adaptive options that are flexible; generate fewer adverse impacts; and help ensure a water-sustainable future.

Section 9

Ethics, Values, and Water — The Challenges of Adaptive Management

Ethics, Values, and Water — The Challenge of Adaptive Management

- *All societies defend water policies on ethical grounds*: what actions are considered good, right, just?
- *Adaptive management*: making decisions based on what we've learned about impacts of previous actions.
- *How are ethics and adaptive management connected*? Both emphasize the need to justify decisions by using knowledge and experience to promote good outcomes.

Adaptive Management — Principles

- An adaptive decision is:
  - *Modest* in scope — imposes a small environmental impact.
  - *Embraces* public concerns — identifies problems before they become urgent.
  - *In reversible* — if it produces adverse outcomes.
- What's a mal-adaptive decision?
  - *Causes conflict* — generates disputes between groups.
  - *Fails to learn from previous mistakes* — thus causes irreversible damage.
  - *Limits participation* — decision-making is confined to an elite.

Three Ways of Thinking about Water Ethics

- *Utilitarianism* — water policy should aim to satisfy the *greatest good for the greatest number of people through*:
  - Economic growth — develop water for power, flood prevention, supply.
  - Efficiency — ensure water is inexpensive and beneficial.
- *Categorical imperative* — water policy should *treat others as we wish to be treated* (e.g., the "golden rule") *by*:
  - Making fair decisions — keep promises and commitments.
  - Putting ourselves in another's' place — would we endorse a policy if it affected us?
- *Stewardship* — water policy should *care for all forms of life. Why?*
  - Rational self-interest.
  - *Because we're commanded* to by higher moral authority.

Applying Utilitarianism — Colorado River

- Challenges:
  - *D*oes everyone benefit equally (i.e., is the "greatest good" allocated fairly)?
  - Is building a dam an adaptive management approach?
- "The farms, cities, and people who live along the many thousands of miles of this river and its tributaries . . . depend upon the conservation, the regulation, and the equitable division of its ever-changing water supply. Distributive works, laws and practices . . . will insure to the millions of people who now dwell in this basin, and the millions of others who will come to dwell here in future generations, a just, safe, and permanent system of water rights. . . . As an unregulated river, the Colorado added little of value to the region this dam serves".

— President Franklin Roosevelt's
Dedication Day Speech at Hoover Dam,
September 30, 1935

Categorical Imperatives — UN Dublin Conference (1992)

- Dublin Statement on Water and Sustainable Development: 170 countries met to discuss need for an *ethically just global water policy* — conclusions?
  - Freshwater is a *finite, vulnerable resource*, essential to life, development, environment.
  - Water development and management should be based on a *participatory approach, involving water users, planners, and policymakers at all levels.*
  - Women play a central role in provision, management, safeguarding of water.
  - Water has an *economic value* in all its competing uses.
- Role of developed countries: (we must) ". . . consider financial requirements and transfers for water-related programs, in accordance with these principles . . . . include realistic targets for implementation ... internal and external resources needed, and means of mobilizing these. These are basic principles of cooperation".

Stewardship — The Legend of the Lost Salmon

The Creator taught the people how to care for this food which was created especially for them. He said, "Do not neglect this food. Be careful that you do not break the rules in taking care of this salmon. Do not take more than you need". He told them if they observed these rules, the salmon would multiply several times over as long as they lived. Legend of the Yakima tribe.

At first the people diligently obeyed the rules, and they lived happily without problems. All along the river there were different bands of people living in their fishing villages, busy catching and drying their supply of salmon.

But one day something strange happened. The people became careless and they neglected to follow the instructions made by the Creator. They became greedy. They did not take care of the salmon. They let them go to waste when they caught more than they needed for their families. They would not listen to the advice from those who were trying to follow the rules. Suddenly the salmon disappeared.

- Challenges:
  - Pacific Northwest salmon may be nearing extinction.
  - Should we protect salmon out of self-interest? (e.g., its economic value, or as an indicator of river health). Or, do we have a *deeper moral obligation?*

Source: US Forest Service — http://www.fs.usda.gov/Internet/FSE_DOCUMENTS/fsbdev2_025634.pdf.

Faith-Based Water Stewardship

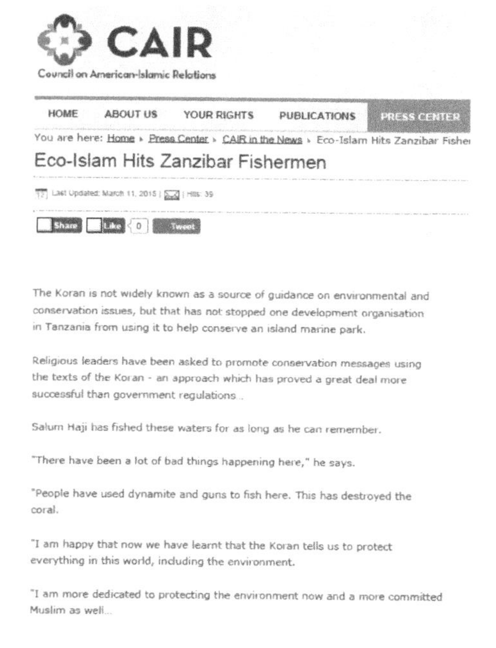

CAIR
Council on American-Islamic Relations

HOME ABOUT US YOUR RIGHTS PUBLICATIONS PRESS CENTER

You are here: Home › Press Center › CAIR in the News › Eco-Islam Hits Zanzibar Fishe

Eco-Islam Hits Zanzibar Fishermen

Last Updated: March 11, 2015 | Hits: 39

Share Like 0 Tweet

The Koran is not widely known as a source of guidance on environmental and conservation issues, but that has not stopped one development organisation in Tanzania from using it to help conserve an island marine park.

Religious leaders have been asked to promote conservation messages using the texts of the Koran - an approach which has proved a great deal more successful than government regulations...

Salum Haji has fished these waters for as long as he can remember.

"There have been a lot of bad things happening here," he says.

"People have used dynamite and guns to fish here. This has destroyed the coral.

"I am happy that now we have learnt that the Koran tells us to protect everything in this world, including the environment.

"I am more dedicated to protecting the environment now and a more committed Muslim as well...

Ecumenical Water Network

The EWN is a network of churches and Christian organizations promoting people's access to water around the world

English Deutsch Français Español

Search Site

About EWN What we do Get involved Resources

Seven Weeks for Water

About EWN

The water crisis

WCC holds photo contest

Do you have a photo that tells a story about water justice? As part of the Lenten campaign "Seven Weeks for Water", the World Council of Churches Ecumenical Water Network (EWN) is promoting a photo contest on Instagram encouraging people to share images of water in their daily lives.

Water justice campaign highlights range of issues, focuses on Palestine this year, says EWN coordinator

The WCC's Seven Weeks for Water Lenten campaign through its ecumenical initiative — Ecumenical Water Network — has gained much attention recently, but the campaign is not new.

Flint's corrosive water sparks debate on US resource inequity

The United States has the biggest economy on planet earth. It is of great concern to Rev. Dr Susan Henry-Crowe, however, that such a moniker for her country does not always mean resources are available to the population in an equitable way.

Quick Links

List of members

Contact EWN

Virtual Water Cube

DONATE NOW

For Discussion

- What are some adaptive solutions that would work for California?
- What's a California example of utilitarianism? Of stewardship?

A Final Adaptive Management Lesson — Trans-Boundary Water Conflict

- Nation-states *must cooperate* to manage and solve water problems — BUT, *process, power, purpose* work differently in the international arena:
  - Nations have unequal power: mistrust, jealousy, rivalry inhibit cooperation and negotiation.
  - No authoritative process for dispute resolution: nations work together to manage problems, making war unlikely (e.g., India/Pakistan; Israel/Palestine) but compliance is entirely voluntary.
  - Divergent goals and purposes: reinforce a lack of trust and a reluctance to share information.

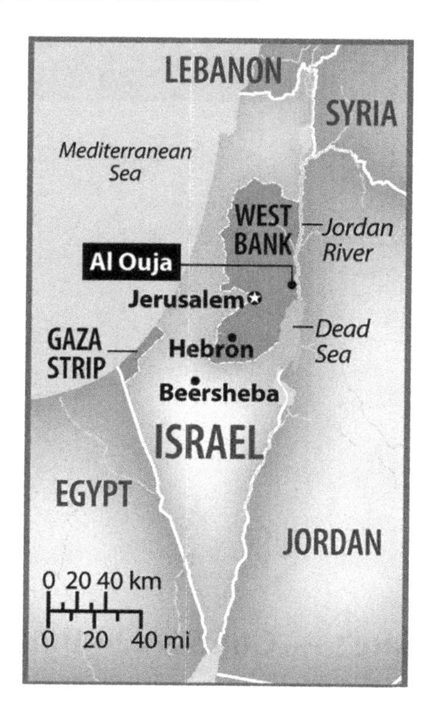

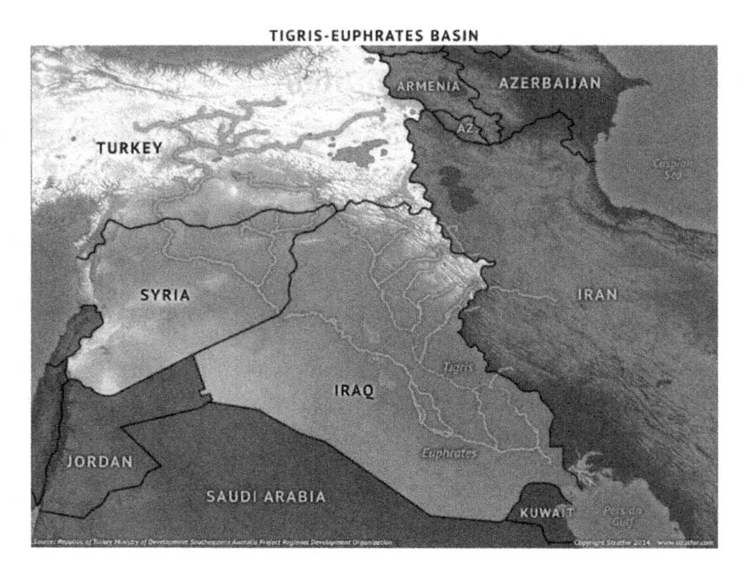

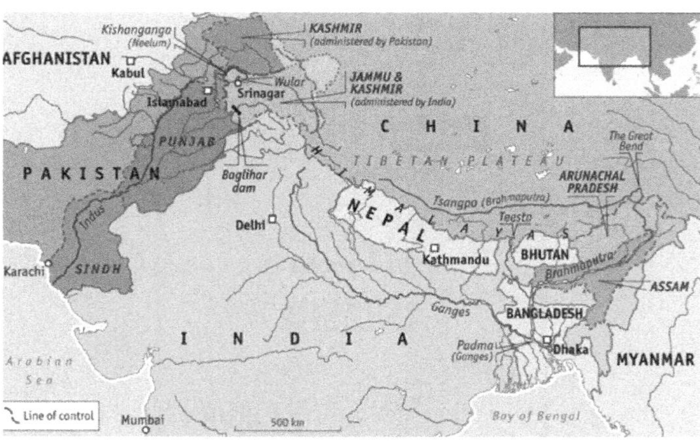

How to Cooperate? — Confidence-Building

- Countries must be assured that if they *comply* with an agreement others will do likewise — and compliance mechanisms must be *explicit in treaties*.
- Agreements must deter parties from cheating — violators will be "caught" (independent verification).
- Agreements must provide tangible benefits — e.g., more water supply, enhanced water quality.

Patterns of Confidence-Building

- *Low cooperation/low confidence*: (Tigris–Euphrates) — Iraq, Syria, Turkey can't agree to jointly manage the basin because they mistrust one another.
- *Partial cooperation/low confidence*: Israel and Palestine; India and Pakistan — parties agree to cooperate because violence will only worsen water conflict.
- *Robust cooperation/high confidence*: (e.g., 1909 US–Canada Boundary Waters) — both sides monitor and report on water quality; seek to protect Great Lakes and other waters; RARE OCCURRENCE.

Tigris-Euphrates Conflict — Low Confidence

- 1980s Turkey began Southern Anatolia Development Project (GAP) to build 21 dams, 19 hydroelectric plants, cultivate 1.7 million hectares of farmland.
    - Projects planned without consulting or accommodating one another — led to sporadic skirmishes between Iraq and Syria, Syria and Turkey.
    - Separatist movements made disputes more intractable — e.g., Syrian support for Kurdish separatists in Iraq.
- 2008 — Iraq, Syria, and Turkey agreed to establish a *Joint Trilateral Committee (JTC)* to manage basin.
    - Iraq and Syria shared a single negotiating position — waters should be allocated equally among three countries.
    - Turkey's response? Allocation should be based on amount of land under cultivation.

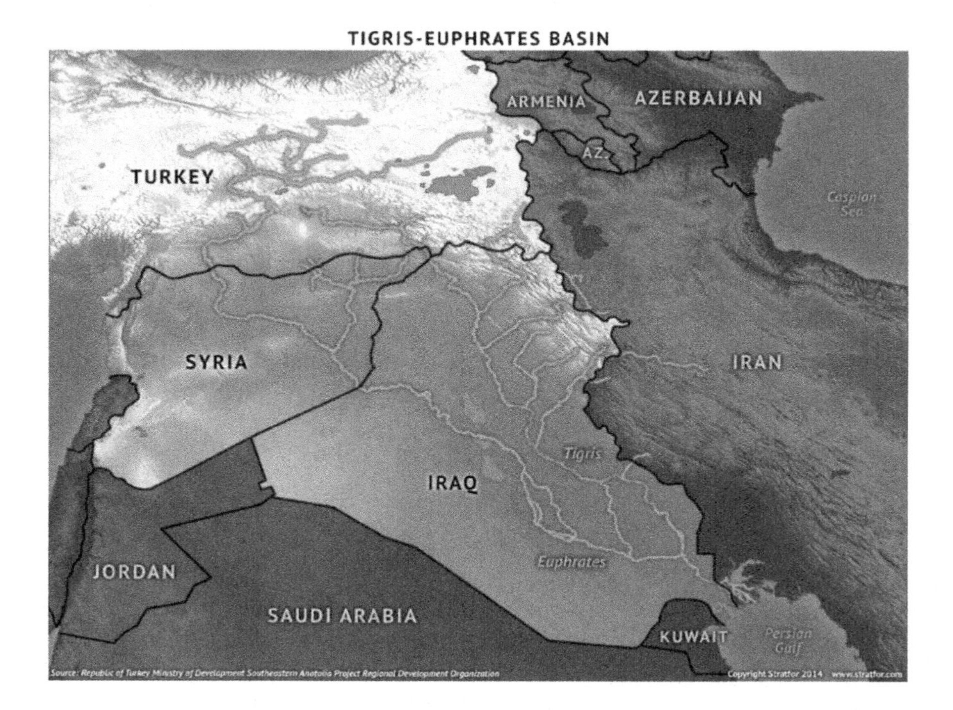

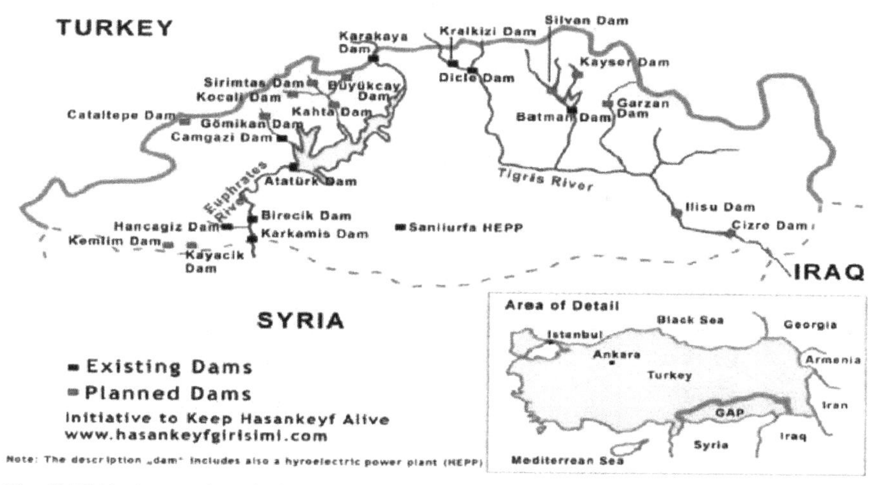

The GAP Project region (including planned and existing dams in 2016). *Source*: Hommes et al., 2016: 10.

...and Today?

- 2008: all three countries agreed in principle to fairly allocate the rivers, engage in prior consultation before undertaking any major projects — Iraq agreed to trade petroleum with Turkey and to help curb Kurdish separatists.
- April 2014: Turkey unilaterally reduced flow of the Euphrates cutting flows to Iraq by 80%, to Syria by 40%, in part to hasten completion of key elements of the GAP project by 2017. Once finished, GAP project will withdraw up to 70% of the flow of the Euphrates — 90% of its flow originates in Turkey.
- https://www.youtube.com/watch?v=8YlGAdm4K7g.

Water as a Weapon? — Iraq

- While countries may not go to war over water, once conflict ensues water infrastructure may become a target.
- Peshmerga soldiers stand guard at the Mosul Dam in northern Iraq on *August 21, 2014*. The Kurdish militia and other Iraqi fighters recaptured the dam from Islamic State jihadists after 12 days of fighting over the crucial dam.

Israel and Palestine — Low Confidence

Susiya — a Jewish West Bank village.

Jordan River flowing into Sea of Galilee.

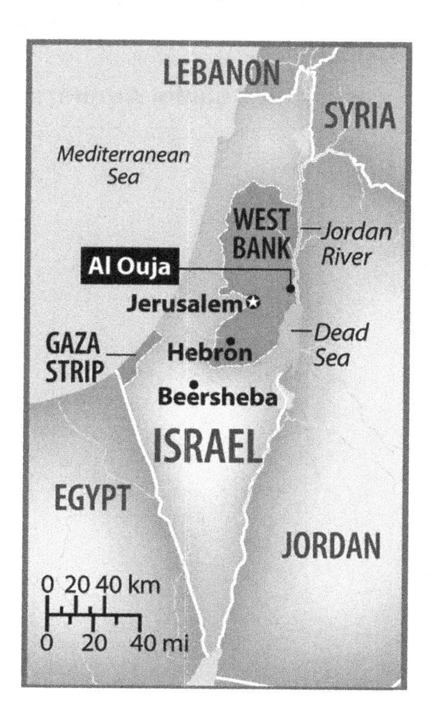

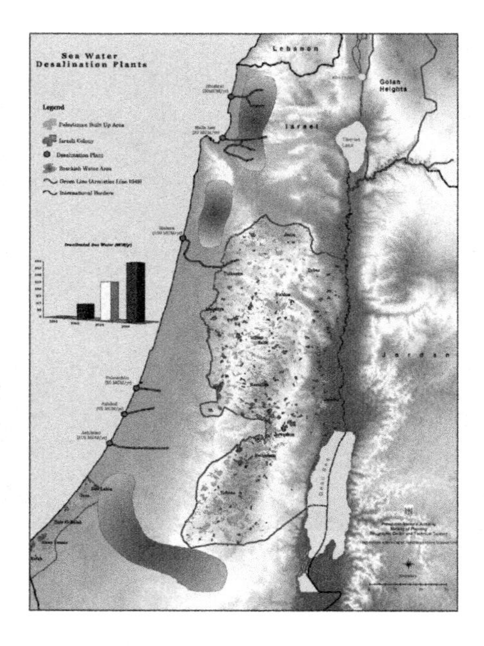

Mountain and Coastal Aquifers

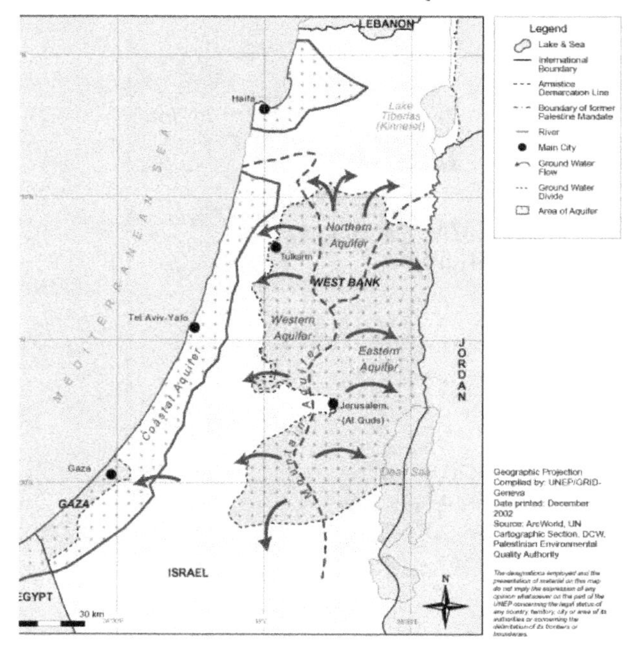

Managing Shared Water Resources

Challenges

- West Bank and Israel share groundwater basins and surface watersheds.
- Water flows toward Israel — which uses 5x as much water per capita as Palestine.
- Surface waters have been fully "appropriated".
- Since 1948, violence has targeted water infrastructure.
- After 1967, Israel took control of West Bank's water and connected settlements to national water network; imposed drastic restrictions on drilling of new wells.

Oslo accord remedies (2001)

- Established a *joint water committee* to "remove ... water and sewer infrastructure from the circle of violence".
- Signatories agreed to manage water and sewer infrastructure and *not to harm or disrupt anyone engaged in operating, maintaining, repairing water works.*
- Palestine given right to extract 20% of annual renewable volume of Mountain Aquifer, draw from Coastal Aquifer as needed.
- Israel would transfer 23.6 mcm/year of water to the West Bank and 5 mcm/year to Gaza strip.

Water and the "Cycle of Violence"

ISRAELI — PALESTINIAN JOINT WATER COMMITTEE
Joint Declaration for Keeping the Water Infrastructure out of the Cycle of Violence

The Israeli and Palestinian sides view the water and waste water sphere as a most important matter and strongly oppose any damage to water and wastewater infrastructure.

The two sides are taking all possible measures to supply water and treat wastewater in the West Bank and Gaza Strip, even in the difficult circumstances of the recent months.

The two sides wish to bring to public attention that the Palestinian and Israeli water and wastewater infrastructure is mostly intertwined and serves both populations. Any damage to such systems will harm both Palestinians and Israelis.

A special effort is being made by the two sides to ensure the water supply to the Palestinian and Israeli cities, towns and villages in the West Bank and Gaza Strip. In order for this effort to succeed, we need the cooperation and support of all the population, both Israeli and Palestinian. We call on the general public not to damage in any way the water infrastructure, including pipelines, pumping stations, drilling equipment, electricity systems and any other related infrastructure.

The two sides also call on those involved in the current crisis not to harm in any way the professional teams that conduct regular maintenance or repair damage and malfunctions to the water and wastewater infrastructure.

Both sides wish to take this opportunity to reiterate their commitment to continued cooperation in the water and wastewater spheres.

Done at the Erez Crossing, This 31 January 2001.
Noah Kinarty — Head of the Israeli side to the JWC
Nabil El-Sherif — Head of the Palestinian side of the JWC

Precarious Cooperation

- Agreement is culmination of a process that began in 1995.
- Provides a means for ensuring *compliance and verification* — each side knows, at local level, how much water is consumed, and for what purpose.
- *However*: negotiators "bracketed out" larger issues–including equitable allocation of total water supplies among the two nations:
  - Re-allocation remains intractable — until resolved, no lasting agreement possible.
  - Inequity remains in national sharing of water and water data — both sides assess problems differently.
  - There is no policy coordination for areas in greatest need (e.g., Gaza), and Jordan River — and little trust in the process of negotiation.

Divergent Views of Conflict

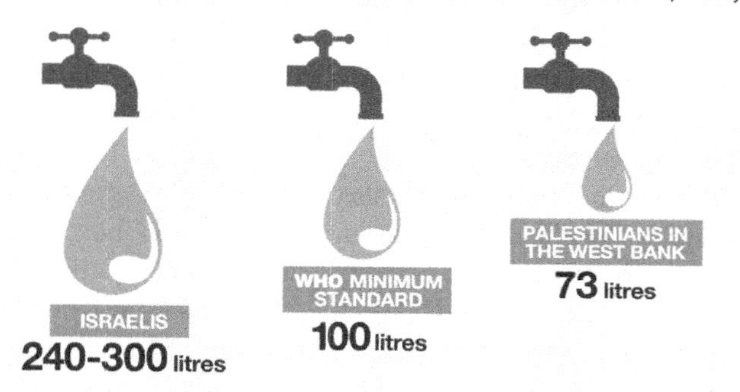

United Nations OCHA, 2019.

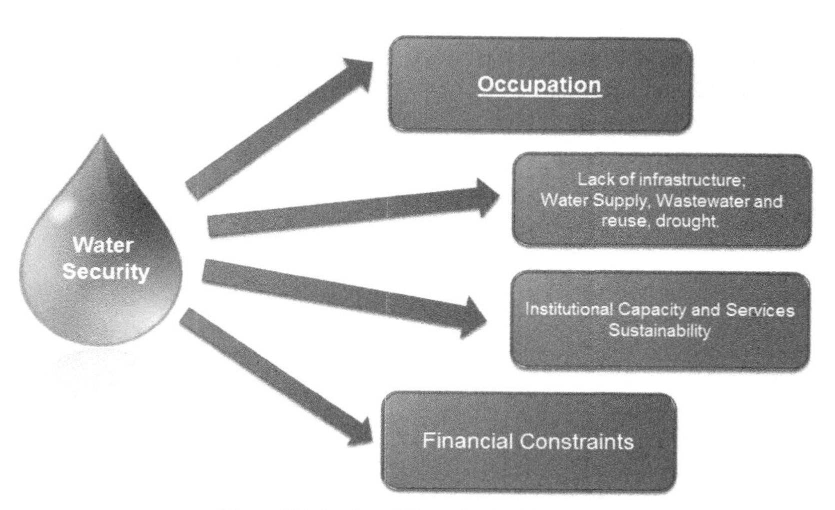

State of Palestine, Water Authority, 2019.

State of Israel, Water Authority: "The Issue of water between Israel and the Palestinians, 2014".

The period of the Interim Agreement was originally determined as five years from the signing . . . However, the two sides have continued to operate according to this Interim Agreement since the time of its signing to the present, even though more than 13 years have elapsed since the agreement was signed. Israel has responded to the needs of the Palestinians and has increased the quantity of water provided to them far beyond that specified in the Interim Agreement.

Israel offered the Palestinians the possibility of erecting a seawater desalination plant in the Hadera area, which would be constructed and operated for them by donor countries, and which would supply water directly to areas in the West Bank. In addition, Israel proposed to the Palestinians the purchase of water for the Gaza Strip directly from the desalination plant at Ashkelon. The Palestinians are well aware of the need to develop a new major source of water (desalination), but are nevertheless not in a hurry to take steps in this direction.

Partial Cooperation — India and Pakistan

- India is building a hydroelectric dam to feed its rapidly growing, power-starved economy — the Kishanganga-Neelum project.
- Pakistan fears India could reduce flows for agriculture — 25% of its economy and employs 50% of population. In May 2010, filed a case with the *International Arbitration Court* to stop it.
- Pakistan also wants to construct its own hydro project — the Neelum-Jhelum project, but is further behind in construction.

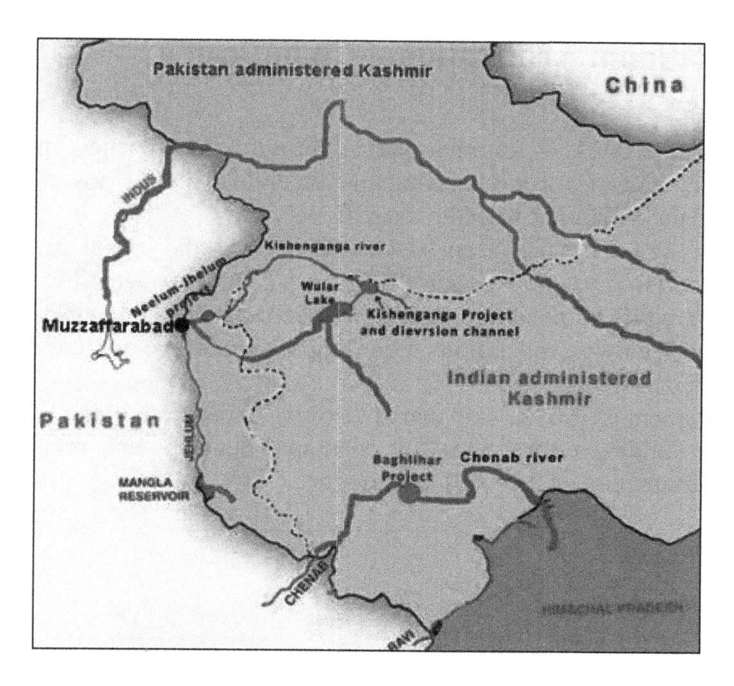
Pakistan administered Kashmir
China
INDUS
Kishenganga river
Neelum-Jhelum project
Wular Lake
Muzzaffarabad
Kishenganga Project and dievrsion channel
Indian administered Kashmir
Pakistan
JEHLUM
Baghlihar Project
Chenab river
MANGLA RESERVOIR
CHENAB
RAVI

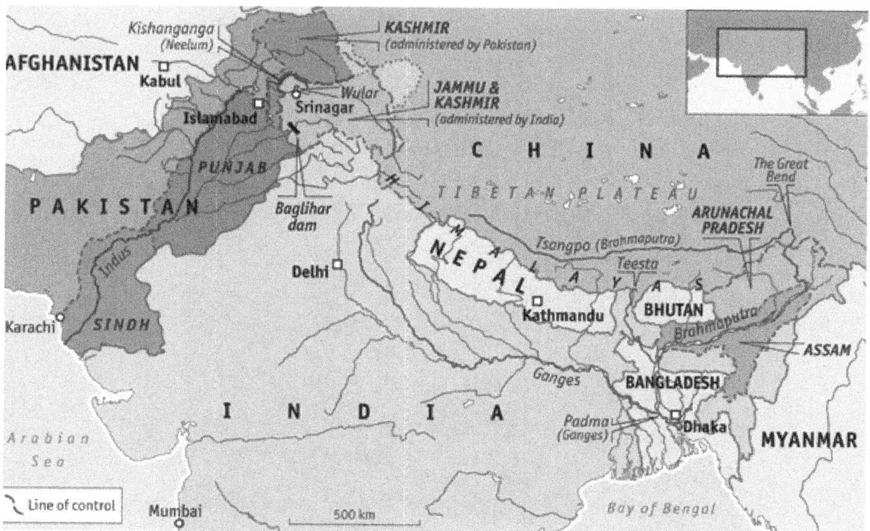
Kishanganga (Neelum)
KASHMIR (administered by Pakistan)
AFGHANISTAN
Kabul
Islamabad
Srinagar
Wular
JAMMU & KASHMIR (administered by India)
CHINA
PUNJAB
TIBETAN PLATEAU
The Great Bend
PAKISTAN
Baglihar dam
ARUNACHAL PRADESH
HIMALAYAS
NEPAL
Tsangpo (Brahmaputra)
Indus
Delhi
Teesta
Kathmandu
BHUTAN
Karachi
SINDH
Brahmaputra
ASSAM
Ganges
BANGLADESH
INDIA
Padma (Ganges)
Dhaka
MYANMAR
Arabian Sea
Line of control
Mumbai
500 km
Bay of Bengal

International Arbitration — What can a "Third Party" Do?

- February 2013 — International Arbitration Court ruled India may divert water from the Kishanganga/Neelum river for power because the dam would not impede water flows.
- Both nations have been instructed to use the *Permanent Indus Commission* to resolve their differences (2020) — agreed to do so.
- India, while allowed to proceed, has been admonished to comply with Indus Treaty's expectations regarding water storage and environmental impacts.
- Settlement is a concession that developing countries will likely increase dam building efforts aimed at economic development, regardless of environmental cost or political friction.

KHEP — 2016 (India) 330 MW.

Neelum-Jhelum project — (Pakistan) 969 MW.

Conclusion — A Way Forward for Water Policy?

- Historically, predominant approach to managing water has been through:
  - Centrally-controlled markets and laws.
  - Large public works that favor growth; disfavor equity and environment.
  - Hierarchical control of rivers and groundwater basins.
- Increasingly, societies aim at adaptive management approaches that rely on:
  - Harnessing local knowledge.
  - Embracing participation and consensus.
  - Seeking *resilience as a goal.*
- However, climate change, population growth, demands for food and energy, and mass migration will continue to be flashpoints for water conflict unless mal-adaptive behaviors can be controlled!

Index

A
adaptive management, 200
adaptive solutions, 166
allocation-based rate structures, 193
American River, 162
Aral Sea, 118
Arizona versus California, 68
Australia, 97

B
Bangladesh, 169
beneficial use, 51
biological oxygen demand, 12
Bolivia, 83
bottled water, 85
Brahmaputra, 169
built environment, 152

C
2014 — California Sustainable Groundwater Management Act, 133
California, 31
California Air Resources Board, 130
Cape Town, 9
Carlsbad project, 185
categorical imperative, 200
Central Arizona Project, 70
Central Valley, 35, 132
Cheonggyecheon River, 146
Chile, 8
cholera, 98
Chumash village, 34
Clean Water Act, 94
Clinton, Hillary Rodham, 15
Coke, 85
Colorado River, 128
Colorado River Compact (1922), 62, 71
compacts, 60
confidence-building, 207
conservation, 21
contaminants of concern, 102
cryptosporidium, 100

D
Dead Sea, 16
degradation, 118
Delaware River Basin Commission, 61
delta smelt, 198
delta tunnels, 183
dendrochronology, 66

Department of State, 15
depletion, 118
desalination, 20
Development Impact Fee, 165
diversion, 118
drought, 152
Dublin Statement on Water and Sustainable Development, 202

E

Edwards, Marc, 108
Endangered Species Act of 1973, 197
environmental justice, 4
European Union, 173
Evian, 85

F

Flint, Michigan, 105
FloodRISE, 167
fracking, 172
Friends of LA River, 140

G

Ganges, 169
gender equity, 7
geo-engineering, 161
Glen Canyon Dam, 196
global demands, 3
Governor Newsom's Water Resilience Portfolio, 53
Gulf of Mexico dead zone, 110

H

Hanna-Attisha, Mona, 108
Hetch Hetchy, 37, 53
Hohokam, 31
Hoover Dam, 202

I

Imperial Valley, 126
India, 205
indirect potable reuse system, 188
International Arbitration Court, 216
International convention on protection of the Rhine, 61
Israel, 16
Israel Water Authority, 185

J

Joint Trilateral Committee (JTC), 208
joint water committee, 213
Jordan, 16, 28
Jordan-Israel Peace Treaty 1994, 17

K

Kishanganga-Neelum project, 216

L

LADWP, 140
levees, 165
Los Angeles Aqueduct, 42
Los Angeles River, 41, 140
low cooperation/low confidence, 207
low-energy water sources, 179
low-intensity conflict, 17

M

Measure W, 95
Mesopotamia, 27
Metropolitan Water District, 78
Mexico, 7
Mexico–US treaty (1944), 71
Millennium Drought (1996–2010), 192
Min River, 26
Monterey peninsula, 20
Morelos Dam, 74
mountain aquifer, 213
Muir, John, 37

Mulholland, William, 43
Murray Darling River Basin, 60

N
Nabataeans, 29
Natural Resources Defense Council, 173
Nile River, 29
non-point pollution, 94
nutrient pollution, 110

O
OCWD Groundwater Replenishment System, 188
Oslo accord remedies (2001), 213
over-drafting, 10
Owens Valley, 43
O'Shaughnessy Dam, 38

P
Pakistan, 205
Palestine, 16
partial cooperation/low confidence, 207
PepsiCo, 85
Per- and Polyfluoroalkyl substances (PFAS), 103
per capita use, 22
Permanent Indus Commission, 218
Perrier-Nestle, 85
Pinchot, Gifford, 37
Point Source Pollution, 94
pollution, 3
post-industrialization, 149
POWER, 53
precautionary policy, 115
prior appropriation, 57
private companies, 78
privatization, 82
PROCESS, 53
Public Trust Doctrine, 52
public utilities, 78
PURPOSE, 53

R
rainwater runoff, 97
Red Sea-Dead Sea canal, 17
reserved rights, 34
resilience, 179
Rhine River, 114
Riparian doctrine, 57
robust cooperation/high confidence, 207
Rodrigo Mundaca, 8
Romans, 30

S
2013 Stream Restoration Agreement, 50
Sacramento Area Flood Control Authority (SAFCA), 162
Sacramento River, 159
Safe Drinking Water Act, 108
Salton Sea, 123
San Francisco, 37
San Joaquin Valley, 131
San Pedro, 49
Seoul, Korea, 146
Sichuan Province, 26
Southern Anatolia Development Project (GAP), 208
Southern California Coastal Water Resource Project, 102
State Water Project, 75
Stewardship, 200
St. Francis Dam, 47
Syria, 16

T
Taihu Lake, 13
Tar-Pamlico River, 112

Tigris-Euphrates, 208
Trans-Boundary Water Conflict, 205

U
1909 US–Canada Boundary Waters, 207
UNESCO, 122
unsightly, 20
urban runoff, 14
US Army Corps of Engineers, 153
US State Department, ix, 15
utilitarianism, 200

W
Water Conservation in Landscaping Act of 2006, 193
water-smart billing systems, 193
water stress, 3
West Bank, 213
World Bank, 122

Z
Zanja Madre, 40

Printed in the United States
by Baker & Taylor Publisher Services